PHEASANT
HEALTH & WELFARE

• D.R. WISE •
M.A., Vet.M.B., Ph.D., M.R.C.V.S.

© D. R. Wise 1993
ISBN 0 9521102 0 2
Cover design by John Williams
Typeset and Printed by
Piggott Printers Limited, Cambridge

Table of Contents

INTRODUCTION
A.	WHY THIS BOOK WAS WRITTEN.	1
B.	HOW TO USE THIS BOOK.	2
C.	OPTIONS FOR A PHEASANT SHOOT.	2
	i) Should one release reared pheasants?	2
	ii) Degree of keepering effort	2
	iii) Stocking strategies	3
	a) Buying poults	
	b) Buying chicks	
	c) Producing chicks	

CHAPTER 1 : EGG PRODUCTION
A.	SELECTION AND ACQUISITION OF STOCK.	5
	i) Size	5
	ii) Plumage	6
	iii) Health	6
	iv) Age	8
	v) Changing bloodlines	8
B.	TREATMENT TO BIRDS BEFORE STOCKING AND TIME OF STOCKING.	9
	i) Brails and wing clipping	9
	ii) Spurs and toenails of cocks	9
	iii) 'Spex'	9
	iv) Time of stocking	10
C.	LAYING FACILITIES.	10
	i) Flockpens	10
	ii) Single harem pens	11
	iii) Indoor systems	12
D.	DIET AND FEEDING.	12
	i) Diet and feeding before the laying period	12
	ii) The breeder diet	13

	iii)	Supplementation of breeder diets with other feeds	14
	iv)	Quantities eaten	14
	v)	Water	16
E.	EGG PRODUCTION PARAMETERS.	16
	i)	The laying period	16
	ii)	Egg numbers	17
F.	EGG COLLECTION, HYGIENE AND GRADING.	17
	i)	Collection	17
	ii)	Hygiene	18
	iii)	Grading	19
G.	EGG STORAGE.	19
H.	PROBLEMS AND THEIR PREVENTION.	20
	i)	Internal parasites	20
	ii)	External parasites	21
	iii)	Bacterial and viral diseases	22
	iv)	Mycoplasma infections	22
	v)	Treading injuries	23
	vi)	Vent pecking	23
	vii)	Poor egg production	24
	viii)	Small egg size	25
I.	STATISTICAL GUIDELINES.	25
J.	RECORD KEEPING.	25

CHAPTER 2 : INCUBATION

A.	HATCHERY DESIGN.	26
B.	SETTING.	26
C.	HATCHING.	29
D.	TAKING OFF THE HATCH.	30
E.	HATCH ANALYSIS.	31
F.	EXPECTED HATCHING RESULTS	33

G.	HATCHING PROBLEMS.		33

CHAPTER 3 : REARING

A.	HOUSING.		36
	i)	General Principles	36
	ii)	Housing Systems and Stocking Densities	37
		a. Moveable huts, night shelters and grass pens	
		b. Permanent brooder house with outside grass runs	
		c. Permanent brooder house with fully covered runs	
		d. Brooder house with no runs	
		e. Mixed systems involving movement of chicks from one unit of accommodation to another	
	iii)	Batch size	40
	iv)	Seasonal stocking rates	40
	v)	Significant design features	42
		a. Flooring	
		b. Corners	
		c. Popholes and doors	
		d. Roofs and ceilings	
		e. Perches	
		f. Handling	
B.	HEATING.		45
	i)	Gas	45
	ii)	Electricity	45
	iii)	Paraffin	46
C.	VENTILATION.		46
D.	LIGHTING.		47
E.	BEDDING.		48
F.	FEEDS AND FEEDING.		49
	i)	Nutrient specifications	49
	ii)	Physical form of feed	50
	iii)	Feed quantity	50
	iv)	Drug inclusions	51
	v)	Feeders	52
G.	WATER AND DRINKERS.		52

H.	**CONTROL OF FEATHER PECKING.**		54
	i)	General	54
	ii)	Light intensity	55
	iii)	Beak trimming	57
	iv)	Bits	58
I.	**DAILY MANAGEMENT.**		60
	i)	Introducing chicks to brooders and the early days	60
	ii)	Hardening off	61
J.	**CATCHING FOR RELEASE.**		62
	i)	General	62
	ii)	Culling	63
	iii)	Trimming or pulling wings	63
	iv)	Poult identification	64
K.	**PROBLEMS.**		65
	i)	The first two weeks	65
		a. Minor mortality during the first three days	
		b. Moderate or heavy mortality during the first three days	
		c. Mortality during the fourth to sixth day	
		d. Moderate or heavy mortality continuing beyond day six	
	ii)	The third week to release	67
		a. Coccidiosis	
		b. Mortality soon after bitting	
		c. Respiratory disease and septicaemia	
		d. Hexamita and Trichomonas	
		e. Worms	
		f. Nutritional problems	
	iii)	Summary of rearing problems	72
L.	**TARGET PERFORMANCE.**		73
	i)	Mortality	73
	ii)	Weight gain	73
	iii)	Economics of poult production	75

CHAPTER 4 : POST RELEASE MANAGEMENT

A.	**THE RELEASE PEN.**		76
	i)	Siting and construction	76

	ii)	Habitat within the pen	76
	iii)	Size of pen in relation to numbers stocked	77
B.	**STOCKING THE PEN.**	78
C.	**AGE AT RELEASE.**	78
D.	**RELEASING COCKS ONLY.**	79
E.	**PROS AND CONS OF BEAK TRIMMING.**	79
F.	**FEEDS AND FEEDING.**	80
	i)	Quantity	80
	ii)	Quality	81
	iii)	Drug inclusions	84
	iv)	Feeding	85
G.	**WATER.**	89
H.	**PROBLEMS AND THEIR PREVENTION.**	90
	i)	Failure to find feed and water	90
	ii)	Exposure and heaping	90
	iii)	Gut impactions	92
	iv)	Tail pulling and vent pecking	92
	v)	Worms	93
	vi)	Protozoal infections	97
		a. Coccidiosis	
		b. Blackhead	
		c. Trichomoniasis and hexamitiasis	
	vii)	Marble spleen disease	99
	viii)	Mycoplasmosis	99
	ix)	Undersized, poorly moulted birds at start of shooting season	100
	x)	Summary of Problems	100

APPENDIX 1 :
ANATOMY 102

APPENDIX 2 :
CONVERSION TABLES 104

Introduction

A. WHY THIS BOOK WAS WRITTEN.

The author decided to write this book as a response to criticism by various members of the Welfare Committee of the British Veterinary Poultry Association. They were of the view that standards of husbandry adopted by some keepers and game farmers are poor relative to those of the more intensive poultry industry and that this leads to unnecessary losses and welfare problems with reared pheasants. The author, a poultry veterinarian and game farmer himself, is not entirely convinced by these criticisms and knows of many estates where husbandry standards are excellent. By the nature of his profession, a veterinarian is bound to see a disproportionate number of problem flocks and to be unaware of the many successful ones.

Nevertheless, it is undeniable that improvements could be made in standards of pheasant husbandry. It is a much more exacting test of stockmanship to produce a pheasant ready for release at six weeks than a broiler chicken for the slaughter house. Many shoot proprietors are completely ignorant of the techniques needed to rear poults and only become interested in pheasants during the shooting season. This can lead to a variety of problems, not least to keepers attempting to rear birds without sufficient facilities to enable them to do a reasonable job. The author has even known keepers who spend their own money to improve their rearing systems because their employers regard extra expenditure as unnecessary. At the opposite extreme, there are a few keepers who exploit their employers' ignorance and sell eggs, chicks or even poults produced at their employers' expense and pocket the money.

It is hoped that this book will provide useful basic information about the standards one should expect for reared pheasant production. As such, it could prove useful to intending or existing shoot proprietors. However, the book is primarily aimed at those who are actually directly responsible for breeding and rearing programmes. The emphasis is placed on the prevention of disease and other welfare problems by the exercise of good husbandry and stockmanship. The author attempts to explain how and why problems occur and how they can be avoided rather than dwelling on precise methods of diagnosis and treatment. There are several different ways of producing fit and healthy pheasants by the start of the shooting season. This text is definitely not a prescription of 'the right way' to do things. There is no single 'right way'. The aim has rather been to bring together technical information on pheasant breeding and rearing which it is hoped will prove useful to both experienced and aspiring keepers.

B. HOW TO USE THIS BOOK.

The text is intended to be used as a manual. Only the most dedicated would consider sitting down and reading it from cover to cover! Use should, therefore, be made of the detailed Table of Contents which will enable the reader to direct himself to those areas that are specific to his particular interest at the time.

C. OPTIONS FOR A PHEASANT SHOOT.

i) Should one release reared pheasants?

Clearly, if one wishes to shoot only parent-reared (wild) pheasants, this book will have no relevance. There are very few successful wild pheasant shoots in the country at present, though it is possible that the number will increase with future changes in agricultural practices. The success of a true wild bird shoot depends essentially upon three factors; suitable habitat, good weather and predator control. A keeper on an estate that is surrounded by unkeepered areas cannot reasonably be expected to have a major impact on the predation of breeding pheasants even though he may kill many predators. It follows that, over most areas of the country, if one wishes to shoot pheasants, one must first release them. The alternative is to shoot those reared by one's neighbour, an unfortunately common practice that is tantamount to theft.

However, this is not to condone a put and take system of shooting. It is merely a recognition that, in current circumstances, shootable surpluses from wild birds alone are unlikely to be generated. It is, therefore, desirable, from a sporting, conservation and aesthetic point of view, to maintain a releasing programme while, at the same time, to give every encouragement to wild bird production.

ii) Degree of keepering effort.

One cannot entirely divorce the subject of shooting from economic realities. A full time keeper whose principal purpose is the provision of sport, as opposed to estate maintenance, can scarcely be justified unless the annual bag exceeds a minimum of 2000 birds. However, as soon as one dispenses with a keeper and attempts to run a shoot on a do-it-yourself or part time basis, one's return rate on released birds generally drops from somewhere in excess of 50 per cent to below 30 per cent. Often, too, the released birds will be more costly because they have been bought in as opposed to home produced. Furthermore, significant production of wild birds is only likely when intense efforts at predator control take place in the spring. This generally implies the presence of a full time keeper.

INTRODUCTION

iii) Stocking strategies.

a) Buying poults.

This is the simplest option for a shoot that has no gamekeeper and, although relatively expensive, is probably the most sensible. Some shooting farmers and others buy in chicks and rear them themselves, but they should realise that almost as much time can be spent in attending the needs of smallish numbers of chicks as large numbers and that they will save little money if they put in the true costs of their own labour. They must also appreciate that rearing facilities and equipment are expensive relative to their limited seasonal use and must, therefore, be prepared to amortise these costs over a long time period.

If a full time gamekeeper is employed, it is extravagant to purchase poults when the keeper could be rearing them from chicks. The rearing season, May to July, does not conflict with the main periods of predator control from January to April and after harvest. A keeper, given reasonable rearing equipment, should have no difficulty in rearing 5000 poults and one would expect a saving over a purchased poult of the labour element plus the game farm profit which, together, would amount to between a quarter and a third of the price.

b) Buying chicks.

This is usually the best option for those estates with a full time keeper. The alternative is to hatch one's own eggs. However, this has several disadvantages with only limited potential for cost savings. Up to date incubation equipment is very expensive, given the short length of the hatching season. Many of the older machines used on estates give unreliable results and are often accommodated in unsatisfactory premises. Even assuming that the equipment is giving good results, the keeper will be hatching a batch of chicks every week for up to ten weeks. It is very much more demanding to rear a large succession of smallish batches than one or two larger batches representing the same overall number of chicks. Stocking the release pens is also much easier if one is not confronted by a multiplicity of age groups.

In general, it is better for those who catch up hens and produce eggs to sell them to game farms and buy back the required quantity of chicks in one or two batches. In round terms, expect to receive one chick from the game farm for every two and a half eggs supplied. If one produces an excess of eggs, the game farm will probably buy them and this will help to offset the costs of the breeder feed required for the laying stock.

c) Producing eggs.

Even though estates should not generally operate their own hatcheries, it usually makes sense for them to catch up hens and produce eggs, given full time keepers. The reasons for this are as follows:

It is bad practice to shoot out all the hens by the end of the season. This implies that one has totally lost interest in production by wild birds and is operating entirely on a put and take basis. However, if one leaves an excessive stock of hens on the ground, the absolute level of predation will be greatly increased because it is density dependent and one will merely suck in more predators. One should, therefore, catch up as many hens as can quickly and conveniently be caught and put them in laying pens. This will still leave some hens uncaught which will have improved chances of breeding success. Ideally, all released poults should be tagged. When catching up in the spring, one could then make it a policy that only tagged birds go Into the laying pens. One would be hoping that untagged pheasants would have been reared by their parents in the wild and that they, in turn would, therefore, be more successful parents themselves. One must accept, however, that untagged birds could merely be reared birds that had strayed from one's neighbour.

The final and obvious advantage of catching up hens and producing eggs is financial. The costs of the stock and the keeper's wages will already have been met. The value of the eggs will be greatly in excess of the extra feed costs. Furthermore, the hens will still be alive at the start of the next shooting season instead of being fox fodder which would have been their most likely fate had they not been caught up.

CHAPTER 1
EGG PRODUCTION

A. SELECTION AND ACQUISITION OF STOCK.

i) Size.

December weights of hen pheasants in the same good body condition range from 850g for wild stock to 1400g for the heavier strains of some game farm stock. Cock weights are usually some 30 per cent greater in both cases, but does weight matter? The answer seems likely to be yes, but we cannot be certain. The purpose of releasing pheasants into the wild is, so far as possible, to restock like with like, yet heavier stock birds will produce heavier progeny since body size is a highly heritable characteristic. It is all too easy for a game farmer keeping back stock for breeding, particularly if selection is made at poult age, to select the best-looking birds from the group. These will usually be the heaviest. There is considerable anecdotal and some scientific evidence to suggest that heavier birds fly less well than lighter ones. This is not surprising since, as weight increases, wing area does not increase at an equivalent rate.

Heavier strains of pheasant may ultimately achieve higher prices from game dealers, but this advantage will be offset to some extent by greater feed intake. Final body size is far more influenced by genetics than by diet. As an example, chicks hatched from caught up fenland hens, maintained to maturity in captivity and fed ad libitum on a pelleted diet, were no heavier than their cousins in the wild. It is, therefore, recommended that potential stock birds are weighed and that the heaviest be eliminated from a future breeding programme. It should be noted that a hen of, say, 1000g at the end of February will normally have gained 150g by point-of-lay. Most of this weight gain, hormonally induced, is a consequence of the development of mature ova in the ovaries and large expansion of the oviduct. It occurs in the three weeks before first egg and is accompanied by an appetite increase of 25 per cent.

ii) Plumage.

Examination of many laying pens will reveal a wide variation of plumage colours in the stock birds from standard through buff, melanistic, speckled and white. This variation is not seen in wild birds and one can reasonably conclude that pheasants with other than normal colour patterns are at a disadvantage in the wild. While some keepers may like mis-coloured birds, the majority of shooters do not. Significant numbers of game farm customers specify that their birds should not be mis-coloured. Should a game farmer be receiving eggs from estates which catch up and retain mis-coloured stock, he may well find himself embarrassed when it comes to selling the chicks or poults from them.

It should be appreciated that plumage colour is highly heritable. White plumage, for example, is a recessive characteristic, probably controlled in pheasants by a single pair of genes. If only one of the pair of genes coding for white plumage is present, the pheasant will have standard plumage colour. If one places a white cock in a breeding pen, half of the progeny of the hens he fertilises which carry a recessive gene for white plumage will themselves be white and the others will be carriers. None of the progeny of the hens not carrying the white gene will be white, but they will all be carriers. Over a few generations, therefore, it is all too easy to end up with a very high proportion of mis-coloured birds. It may be reasonable for keepers to want a few odd coloured birds in their release pens as marker birds and it will add to their interest. However, such birds should never be put into breeding pens the next season, particularly when flock mating is practised.

iii) Health.

It is obvious that one should only select stock in good body condition. However, this is itself not a guarantee that the selected birds are not carrying disease. There are three diseases about which one should primarily be concerned:

a) Salmonellosis.

If batches of poults reared the previous season have suffered from clinical salmonellosis in early life, but have not experienced excessive mortality, it is probably reasonable and expedient to release them. However, since they may well remain carriers of the bacteria, it is sensible to tag them and, when catching up the following season, to reject such birds as future breeding stock.

b) Mycoplasmosis.

This is a chronic disease with a very prolonged carrier state. It may be transmitted through eggs albeit only at a low rate. Many birds may be carriers with no clinical signs. Those that are clinically affected may cough but, more often, there is a slight swelling below and around the eyes; the birds may rub their eyes and the eyelids may close over *(fig 1)*. The disease appears to be on the increase in pheasants. It was probably originally introduced into

Fig. 1
Swelling of the infraorbital sinus of a chicken caused by **Mycoplasma gallisepticum**. In pheasants, the sinuses tend not to bulge outwards so much, but the swellings cause the eyelids to close.

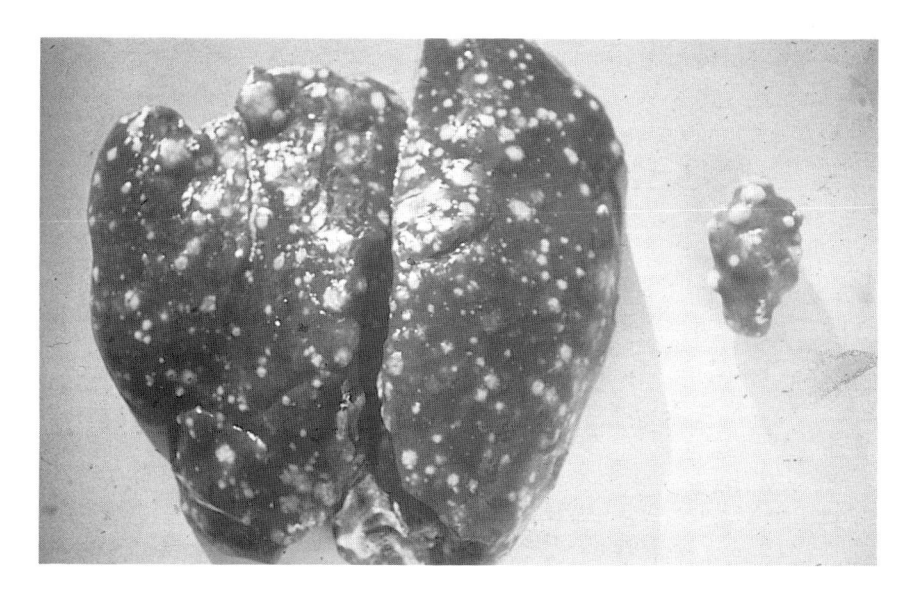

Fig. 2
Granulomatous lesions in the liver and spleen caused by avian tuberculosis. Such lesions are typically seen in adult pheasants which are usually emaciated.

pheasants from infected broody hens or bantams. Obviously, clinically affected birds should be rejected as stock birds and, preferably, killed. If the problem occurs in a batch of poults and it is decided subsequently to release them, such birds should be tagged as recommended in the case of salmonellosis. A rapid blood test is available for mycoplasma via a veterinarian which could allow the detection and elimination of carrier birds the following season. Alternatively, all tagged birds could be rejected as future breeders.

c) Avian Tuberculosis.

This is not transmitted through eggs. It is chronic and usually takes a long time to lead to clinical signs. These tend to be non-specific and, ultimately, lead to emaciation. Because of the long duration of the disease, first season breeders rarely suffer. If birds are retained on a game farm for a second season or, alternatively, if a keeper catches up second season stock, tuberculosis may well lead to poor performance and significant mortality during the laying period *(fig 2)*. In free range conditions, the infection cannot be avoided since it is widely present in other wild birds such as pigeons and starlings. It tends to be in the later stages of the disease that the bacteria are rapidly released from infected birds resulting in heavy contamination of the ground. The bacteria are very tough and persist in the environment for long periods. A keeper can go some way to avoid second season stock when catching up if he has a comprehensive tagging programme at release. Should more than a few birds die of tuberculosis in a laying pen in one season, it would be sensible to resite the pen before a subsequent season.

iv) Age.

When game farmers select poults for future breeding stock, they have the option of selecting from specific hatch dates. There is considerable evidence to suggest that late hatched birds - those hatched after the end of June -make less satisfactory breeding stock. While May hatched pheasants appear in the following year to lay as well as those hatched in June, they do have to be fed for an extra month and this will add approximately 17p per hen in feed bills.

If a game farmer does decide to hold stock birds over for a second season, having satisfied himself that there is no tuberculosis problem, he can expect approximately ten per cent less eggs from the hens than in the first year. However, the eggs will be slightly bigger and production usually starts a few days earlier.

v) Changing bloodlines.

It is common to encounter keepers and landowners who wish to buy in 'new blood', presumably fearing that their birds will become inbred. Given that one uses at least ten cocks and one hundred hens as stock and mates randomly, the coefficient of inbreeding will be very low and it will be many years, if ever, before inbreeding becomes a problem.

However, it may be argued that, on many estates, the pheasants are not randomly selected in the first place. If one constantly breeds from hens caught up at the end of the season, one is likely to be selecting those birds which have escaped shooting by dint of being poor fliers. Under such circumstances, an occasional infusion of 'new blood' may be desirable. There would be no point, however, in obtaining stock birds from another estate with a similar selection policy.

B. TREATMENTS TO BIRDS BEFORE STOCKING AND TIME OF STOCKING.

i) Brails and wing clipping.

Before placing stock birds into open topped laying pens, some means of keeping them in must be used. Even in pens which are roofed, there is something to be said for preventing full flight to avoid damage to birds on roof netting. One of two methods may be used. One wing can be tied in a flexed position by use of a proprietary brail or with brailing tape. Alternatively, the primary flight feathers of one wing can be clipped. New feathers will not grow until September so the birds will be effectively grounded until this date.

The advantages of wing clipping are that the process is quicker and cheaper than brailing. The birds seldom attempt flight and tend to panic less than those in brails which seem never to come to terms with their degree of restraint. The disadvantage is that wing clipped birds, released from their laying pens in mid June are more vulnerable to predation until their new feathers have grown out some months later.

ii) Spurs and toenails of cocks.

In order to reduce potential damage to hens through treading injuries, one should blunt the spurs and nails of the inner toes of the cocks. This is usually done with the hot blade of an electrical debeaking machine. In effecting this operation, do not remove too much, but concentrate on leaving rounded edges to the spurs and nails or one may do more harm than good.

iii) 'Spex'.

These opaque plastic devices clip into the nostrils and act as barriers to forward vision. They are highly effective in preventing an array of vices such as cannibalism, vent pecking and egg eating. Even when applied to cocks as well as hens, they have no adverse effects on fertility.

Many people find 'spex' to be distasteful and consider that they should be condemned on welfare grounds. One should realise, however, that birds such as pheasants with eyes placed at the sides of their heads, have much wider fields of vision than do humans, but have little in the way of binocular vision. 'Spex', therefore, are likely to be far less inhibiting to pheasants than equivalent devices would be to ourselves. One could take the view that one should provide such a good habitat in the laying pen that vices will not occur and 'spex' would thus become unnecessary. The author has yet to see such a pen. Were one to exist, it would probably contain so much cover that egg collection would become extremely time consuming and uneconomic. This is not an argument to promote 'spex' in place of a decent habitat. It is certainly feasible to provide laying quarters which minimise vices even if not eliminating them altogether. One must then make the personal choice as to whether it is more acceptable for a small percentage of one's stock to suffer from vices rather than inconveniencing all of the birds by fitting 'spex'.

iv) Time of stocking.

Regardless of the type of laying pen used, birds should be placed in laying pens no later than the end of February. With the typical outdoor systems, birds should certainly not be placed in them too soon because the ground will become fouled. With the single harem system, it may be advantageous to stock each unit with a cock slightly before the hens are introduced since this mimics the natural situation. Simultaneous stocking of both sexes, however, is more normal and produces satisfactory results.

C. LAYING FACILITIES.

i) Flock Pens.

The most usual laying pen currently encountered in the U.K. is the open-topped, flock pen containing anything from two dozen to several hundred hens. Normal cock to hen ratios vary from 1:6 to 1:8. These pens are usually semi-permanent and obviously vary greatly in size. As a rule of thumb, some 4 to 5 square metres (40-50 square feet) of area should be allowed per bird stocked. Some pens are stocked more heavily than this but, generally, egg production and fertility are better when the space allowance is generous. Pens should be sited in such a way as to provide maximum wind protection and minimum outside disturbance. The pens should only be used from February to June and, thereafter, closed in the interests of disease control. Some laying pens, for security reasons, are sited within large release pens. At the end of the laying season, the wire is often rolled up from the sides of such laying pens, giving released poults access to the area later in the season. This type of practice should be avoided.

Laying performance is generally superior when flock size does not exceed 150 birds/pen. In other words, better results are to be expected when a given area is divided into four pens of 150, than when a single pen for 600 birds is provided.

Bare, open pens enable all cocks to see each other simultaneously, resulting in fighting, stress of hens, poor egg production and infertility. It is desirable, therefore, to 'baffle' the pen into sub-units, providing cock 'territories'. Each 'territory' should contain feed and water. Low perches should be available. (If the pen is netted over and the stock birds left with full wings, perches can obviously be higher.) Access to perches will provide hens with some respite from the constant attentions of the cocks.

Pens should also be provided with adequate numbers of, for want of a better description, nest boxes. Pheasants will seldom lay in typical, hygienically designed nest boxes such as provided for chickens. However, they will seek out secluded places. Sometimes these merely consist of cut fir branches, but these are less than ideal as it should be one's aim to prevent the eggs from becoming wet and muddy between the time of lay and collection. Often, cut-down plastic barrels or old wooden broody coops will be selected as nests, particularly if camouflaged with branches. These should be sited on ground not subject to waterlogging. It is often worthwhile to make a circular depression in the soil in the nest boxes as this can reduce cracked eggs. The nest boxes can contain good quality, mould-free straw (not hay) if desired, provided that it is changed regularly and does not become wet. This, too, will reduce problems with cracked eggs and with egg eating as the eggs tend to drop through the straw. Many texts condemn the use of straw in nest boxes due to the risk of aspergillosis. It is the author's personal view that this risk has been over-emphasised. For over a quarter of a century of game farming he has been bedding his nests with straw and has yet to experience any associated problems. Wood shavings, although commonly used in nests of chickens and turkeys, do not allow the eggs to penetrate and may actually encourage egg eating.

Often, eggs will be concentrated in one or two nests to the exclusion of others. This should be discouraged by providing more nests or by moving and rearranging those not used. Excessive numbers of eggs in one nest will not only result in more cracks, but also the continued development of the eggs after they have been laid since a succession of hens will continue to heat them up. This will adversely affect hatchability. The presence of broody hens, usually late in the season, is particularly harmful in this regard and such birds should be removed from the pen.

ii) Single harem pens.

Less common than in the past, such pens contain a single cock and an associated harem of approximately eight hens. The pens themselves may either be large and static, providing the same sort of area per bird as recommended for flock pens or they may be quite small and only provide some 0.5 sq.m (5sq.ft.) per bird. In the former case, the collector must enter the pen to collect the eggs. In the latter case, eggs are collected from the outside and the pen itself must be regularly moved throughout the season on to clean ground.

Egg production tends to start earlier in harem pens and egg production overall may be up to one quarter better than in flock pens. This is probably due to the lower stress levels on the hens which avoid the constant harassment by many different cocks. Nevertheless, cannibalism can sometimes be a problem in the smaller single harem pens. This cannot readily be prevented without recourse to 'spex' or beak trimming.

The advantages of more eggs per hen normally achieved in single harem pens tend to be partly offset by poorer fertility. In an attempt to improve this potential problem, early eggs can be identified to pen of origin by indelible marker and subsequently candled after approximately five days of incubation to check for fertility. In this way, substandard cocks can be identified quite early in the season and replaced.

The single harem system has tended to go out of favour because it is more expensive to establish and considerably more labour intensive than the flock pen. Furthermore, the best designed and managed flock pens achieve as many eggs/hen as the average group of single harem pens and the eggs are usually more fertile.

iii) Indoor system.

It is quite feasible for game farmers to establish flock pens within pole sheds and to stock at much greater densities than can be achieved outside. Birds can be stocked at poult age and kept within the pens till the end of lay. Furthermore, a pole shed system makes it easier to provide artificially extended day length to stimulate earlier egg production which will give ten to fifteen per cent more eggs by mid June.

This system is probably not suitable for housing stock that are caught up at the end of the shooting season as it will be too alien for them. Equally, birds from such a system should not be released on to an estate at the end of the laying season for the same reason.

In competent hands, good results can be obtained from this system with extremely low overall losses from six weeks to end of lay. However, to achieve this level of performance, permanent beak trimming, 'spex' and, preferably, saddles to protect the hens' backs are necessary as well as a good standard of stockmanship. In less skilled hands and in the absence of the artificial aids listed above, which are shunned by many, there can be unacceptably high losses through vices.

D. DIET AND FEEDING.

i) Diet and feeding before the laying period.

Pheasants should be placed on a breeding diet at some stage between two weeks before anticipated point-of-lay and point-of-lay itself. If specially formulated pre-breeder or maintenance pellets are fed before this, then introduce breeder feed at point-of-lay. If only grain is fed, introduce the breeder diet two weeks sooner.

Wheat alone is an unsatisfactory source of nutrition for birds over winter. Although it will contain adequate energy and protein, it will be lacking in certain minerals, vitamins and essential fatty acids. Wheat, as a supplement to natural food available to birds before they are caught up, is probably quite satisfactory although, with its high energy to protein ratio, it predisposes to excessive fatness when given *ad libitum*. It should be realised that the natural food available to future breeding stock penned in winter quarters is often non-existent. Birds penned in such a manner are better fed with maintenance pellets all the time or, if this is considered too expensive, with pellets for one week in four and grain for three weeks in four. Oats or barley, though less favoured by pheasants, are probably better for them at this stage than is wheat. Maize should definitely be avoided because of its very high energy level.

ii) The breeder diet.

Many national and local compounders produce pheasant breeding diets. Such diets are invariably pelleted. It should be appreciated that breeder diets for chickens are not the same as layer diets. The latter contain less vitamin supplementation and, while they will support optimum egg production, will not allow good hatchability. Some pheasant keepers, in the interests of economy, attempt to feed chicken layer diets with disastrous consequences. Even chicken breeder diets are not optimum for pheasants in that they tend to contain an excess of calcium. One should, therefore, select a breeder diet specifically formulated for pheasants.

The choice of compounder should be based either on experience or the personal recommendations of other users. Pheasant breeder diets from different sources tend to vary very widely in price with the most expensive often being 15 per cent more costly than the least. It should not necessarily be assumed that the most expensive is the best. The reason for variation is more often to be found in differences in advertising and promotion costs, or by the size of backhanders that some company representatives offer to keepers. Generally speaking, it is wiser to buy from a compounder whose mill makes significant tonnages of poultry diets for stock other than game. The machinery required to mix and manufacture diets appropriate for poultry is generally more sophisticated than that needed, for example, for pig and cattle diets. One should, therefore, be somewhat suspicious of a compounder who only makes game diets and not other poultry feed.

Breeder diets should not routinely contain any drugs except, possibly, one to prevent blackhead (*Histomonas meleagridis* infection). Such a drug should also protect against Trichomonas and Hexamita infections (see later).

Ideally, breeder diets should be fed within six weeks of manufacture and should be stored in cool, dry conditions. Under no circumstances should one buy from a compounder who endeavours to deliver one's entire breeder feed requirements in one drop at the start of the season. Two drops should be regarded as the minimum.

Feed problems that have serious adverse effects on production or hatchability are, fortunately, rare. When they do occur, however, they are often extremely difficult to pin down, particularly those affecting hatchability, because they may not become obvious until several weeks after feeding by which stage the suspect feed may all have been eaten. For this reason, it is theoretically sensible to take and retain samples of each feed consignment delivered. To take a representative sample, it is necessary to add a handful of feed from each of several sacks and to mix and store in a polythene bag in a cool, dark place.

iii) Supplementation of breeder diets with other feeds.

a) Grain.
This should not be fed in significant quantity as it will cause an imbalance of nutrients in the diet. If keepers insist on feeding it, it should be limited to a scratch feed of no more than 3g/bird/day and be regarded as entertainment for the birds.

b) Greenstuff.
Many keepers cut and offer various vegetable materials to their laying pheasants, particularly when grass has been eaten from the pen. Such material will add little in the way of essential nutrients, but may keep the birds happy and so reduce egg eating, fighting and general stress.

c) Insoluble grit.
Birds that eat pellets and nothing but pellets do not really need insoluble grit in their gizzards as an aid to digestion. However, occasional pheasants develop semi-impacted gizzards due to ingestion of long fibrous plant material. Even if this doesn't make them ill, it may reduce appetite and, thus, egg production. In cases where the pen contains no natural grit, it is beneficial to scatter flint or quartz over the food once a fortnight at the rate of approximately 0.5kg (1lb)/100 birds.

d) Soluble grit.
When a well-balanced breeder diet is provided, there is no need for oyster shell or limestone grit, the sole purpose of which is to provide a source of calcium for eggshell formation. If offered in a separate feed hopper experience suggests that very little is consumed. However, were it to be consumed, it would normally be expected to have adverse consequences. It is, therefore, better not to offer it at all.

iv) Quantities eaten.
The amount of feed eaten will depend upon the energy content of the diet, the size of bird, its activity and rate of lay, and upon the environmental temperature.

EGG PRODUCTION

If one considers the typical feed intake of a fairly standard hen pheasant of, say 1100g, in a typical season, it will probably be eating approximately 60g/day of maintenance pellets or 55g/day of wheat some four weeks pre point-of-lay. In the three week run up to lay, feed intake will increase to approximately 75g/day of breeder pellets and body weight will have reached 1250g at point-of-lay. Once egg production starts, feed intake drops back to approximately 70g/day, but builds up again after a fortnight and thereafter increases to 80g/day. Then weight tends to remain reasonably constant through lay but, by the end, there is less weight of reproductive material in the hen and more fat. The example given is based on an experiment involving ad libitum fed birds in a pole shed. It can be argued that such a regime may produce birds that are too fat. Experience with other poultry suggests that excessive fatness does not interfere with egg production, but does lead to the production of eggs with less than optimum fertility. Cocks probably consume about 75g/day and tend to lose significant quantities of weight during the breeding period.

It is important to know the feed consumption per bird of one's laying pen even when *ad libitum* feeding through hoppers. In many communal pens, there are inadequate numbers of hoppers and, although feed may be available at all times, pecking orders and territorial aggression may lead to suboptimal feed intakes among both hens and cocks. If among the former, egg production will suffer. If cocks are not all getting a fair share of feed, fertility will be poor. Disappointing performance for this reason is very common and easily corrected. One feed hopper per fifteen hens is a satisfactory number.

For those who hand feed, extreme attention to detail is necessary. In theory, if one feeds twice a day an amount that will be completely eaten in approximately half an hour, feed consumption per bird will be in the order of five per cent less than that of an *ad libitum* fed bird. This will probably produce a fitter, less fat bird which lays equivalent numbers of eggs, but whose eggs are more fertile. This is probably the extent of the 'upside potential'. The 'downside' is much greater. It is essential to ensure that each bird gets its fair share of food. As less eating time is available, the feed must be spread widely to avoid excessive competition. This often involves scattering it on the ground. In rain, some may be spoilt. Starlings and other small birds may eat significant quantities and the constant pecking about on the ground increases the probability of parasitic infection. The best advice, therefore, is to recommend that those who get good results through hand feeding should stick to their method. Others should either feed *ad libitum* through hoppers or attempt hand feeding with extreme care and monitor feed intake and egg production very carefully.

Environmental temperature affects both egg production and feed intake. In very cold conditions, appetite will be greater, but, notwithstanding, egg production is usually suboptimal. In hot conditions, feed intake falls and, with it, egg production. A heatwave towards the end of May can push hens out of lay prematurely. For every 1°C rise in temperature from 1°C to 30°C, appetite falls by approximately 1.5%. This effect is not linear and the drop is proportionately greater at the higher temperatures. Under such circumstances, hens will not be eating enough to obtain sufficient nutrients for optimum production.

v) Water.

When in lay, a hen pheasant, eating 75g of pellets a day, may be drinking in excess of 200ml (one third of a pint) of water. It is, therefore, just as important to site plenty of drinkers in laying pens as plenty of feeders because the same comments relating to peck order and dominance apply. Because water is present in the pen, this does not imply that one's management obligations have been fulfilled. There must be a multiplicity of well-spaced drinking points.

It is important that drinkers be kept clean. Warm, dirty water rapidly becomes foul and multiplies bacteria. It is a potential disease hazard but, worse, it will result in suboptimal water intake and hence poor egg production. Even when water is relatively clean, containers may be rusty. Birds have a very limited range of taste, but they can detect rusty water and are repelled by it.

When cleaning drinkers, it is common practice to tip the dirty water on to the ground in the laying pen. This is extremely bad management. The ground becomes muddy, as do the feet of the hens, and hence the eggs become more filthy than they need otherwise be. Also, an ideal breeding area for bacterial and parasitic disease is created.

E. EGG PRODUCTION PARAMETERS.

i) The Laying Period.

Egg production usually starts in the first week of April. Point-of-lay varies somewhat from year to year and from region to region. One should expect 50 per cent egg production some two weeks later and peak egg production approximately three and a half weeks after start of lay. Good flocks will stay at peak production for up to six weeks. Thereafter, levels of production will start to decline.

The end of the laying period is not determined by the cessation of egg production, but, rather, by the time of year. Eggs laid after June 20th will not convert to poults ready to release by the end of August which can reasonably be regarded as the latest date on which one should consider release. Some keepers like to let their stock birds out of the laying pens earlier so that they have the chance to lay and hatch a clutch of eggs in the wild. It should be realised that, if the chicks so hatched are to stand much chance of survival, their mothers should be released at the latest by the end of the third week in May. At best, each hen released in this way is unlikely to produce more than three fledged poults. Had she remained in the laying pen until mid June, one might anticipate that she would lay a further dozen eggs which would convert to eight poults.

ii) Egg Numbers.

There is enormous variation in egg production between flocks and between seasons. A cold start to the season, while not necessarily affecting the date at the start of lay, will cause a slower rise in production to peak numbers. Very hot weather in late May and early June is often responsible for a rapid decline in egg numbers late in the season.

In general, those flocks with high production peaks tend to produce more eggs overall. In typical game farm flocks maintained in communal pens, peak levels of production are usually 70 - 75 per cent. In other words, a farmer starting the season with a thousand hens would expect to be collecting 700 to 750 eggs per day over a period of a few weeks in mid season. The total of eggs collected from such a farm to mid June would be 35 to 40 per hen over a period of approximately 75 days. The figures quoted are average - not good. Many egg producers using flock pens, particularly those with fewer hens and less densely stocked pens, achieve substantially better results with peak production figures of 80 - 90 per cent and total eggs/hen to mid June of up to 50. Such results indicate an enormous potential for improvement by the average producer and suggest that most laying pens are less than ideal in design.

F. EGG COLLECTION, HYGIENE AND GRADING.

i) Collection

The recommendations for poultry hatching eggs are that they should be collected five to six times a day and that only eggs that are laid in nest boxes be incubated. Floor laid eggs should be rejected. Such advice would clearly be totally impractical when applied to pheasants. Nevertheless, every effort should be made to collect clean, dry eggs and to avoid the situation arising in which collectors find large numbers of eggs in one nest. It is, therefore, necessary to provide plenty of nests which the hens will actually use, to protect these from rain, to keep them clean and to collect eggs at least twice a day.

The time of the final collection is very important. Most pheasant eggs are laid in the afternoon. Ideally, the final egg collection should not be before 6 p.m. If it takes place too soon, eggs will be laid after the final collection and will remain in the nests for up to 20 hours. They may be frosted overnight in cold early season conditions and will certainly be sat on by a succession of hens the following day which will subject them to a period of heating up. Both eventualities will impair hatchability.

Eggs are probably best collected in plastic coated wire baskets which should not be overfilled as there is a tendency for eggs at the bottom of the basket to be cracked in such circumstances. It is not a good idea to put hay or straw in the bottom of the basket. Ideally, one should collect clean and dirty eggs separately and process them separately, but such advice is not always practicable.

ii) Hygiene.

Once eggs are collected, they should be cleaned as soon thereafter as possible. Formaldehyde fumigation is often used to sterilise the shells of poultry eggs. However, this technique is not suitable for eggs contaminated with faeces or mud and is, therefore, not to be recommended for pheasants. The best method is to wash the eggs in a detergent/disinfectant solution. There are several proprietary compounds on the market for this purpose, usually based on hypochlorite or quaternary ammonium disinfectants.

When an egg is laid, it immediately starts to cool and the internal contents tend to shrink, creating a suction at the surface. If the shell is contaminated with bacteria, which it will be - and very heavily so in the case of visibly dirty eggs - some of the bacteria will be sucked through any open pores in the shell. The situation is much worse when the egg is wet. Once the bacteria - or moulds - are through the shell, they tend to remain on the outer shell membrane just inside the shell for approximately 24 hours. Some may then penetrate the outer and inner shell membranes and get into the egg proper, causing it to rot or the embryo to die. Egg white contains a substance known as avidin which binds iron. Bacteria need iron to multiply. Egg white thus has a defence against bacteria. It is extremely important not to wash eggs in water containing iron such as might be expected in the presence of rust because this will overcome the avidin defence. By no means all the bacteria reaching the outer shell membrane manage to penetrate it. Some will remain on it throughout incubation and may infect the chick when it hatches either through the navel or the respiratory tract.

Properly conducted egg washing will kill bacteria on the shell surface and may destroy a proportion of those that are in open pores or on the outer membrane. Manufacturers recommendations should be followed when washing with the proviso that, when eggs are very dirty, washing solutions should be changed more frequently than recommended. It is important for the solution to be warmer than the eggs to be placed in it, otherwise dirt may be sucked in rather than washed off. However, temperatures in excess of 115°F will kill the embryos. Electrically heated, thermostatically controlled buckets are often used to hold the washing solution. These may be placed on rotating turntables so that, when a basket of eggs is placed in a bucket, it will result in agitation of the eggs and assist the cleaning process. Eggs should not remain in the washing solution for much longer than three minutes. The basket can then be left to drain. Eggs that remain visibly dirty can be dipped back in the solution and washed by hand. They should not be rubbed with cloths as this may force contaminated material through the pores of the shell.

Certain operators develop allergies to egg washing solutions. In such cases, rubber gloves should be worn. Unfortunately, slippery eggs coming out of solution will be more likely to be dropped when gloves are worn.

After eggs have been washed and graded (see below), they should be left to dry before either being loaded on to cardboard egg flats or being placed in store. If one is using cardboard flats, it is important that they be clean, preferably new.

Eggs should be handled as little as possible after they have been washed and then only by operators with scrupulously clean hands.

iii) Grading.

Clearly, not all eggs that are collected are settable. The usual difference between eggs collected and those set is in the region of 8 per cent. No eggs that are cracked or have been punctured by toenails should be set. Obvious cracks should be discarded before washing. Normal egg weights vary from approximately 27g to 36g. Early season eggs tend to be smaller as do those from small strains of stock birds. Eggs that are significantly smaller than 25g are unlikely to hatch and, if they do, will produce chicks of poor viability. They should not be set. Very large eggs are often double yolked and these, again, will not hatch so should not be set. A proportion of eggs laid will have shells totally or partially covered with a chalky deposit which gives a rough feel to the surface. The abnormality is not visibly obvious in wet eggs which have just been removed from the washing fluid, but becomes so as the eggs dry. Such eggs are known as 'held' eggs because they have been retained in the hens' shell glands for too long. A second shell has started to be deposited on the surface of the first. If one scrapes the chalky deposit with a sharp knife, a shell of normal appearance is revealed. 'Held' eggs are seldom picked up from nests, but are usually found lying in the open, often in the morning. Because such eggs continue to develop for too long within the hen, they are past the stage at which storage becomes possible by the time they are laid. They thus never hatch. The presence of large numbers of 'held' eggs suggests that hens are being stressed, harassed or generally disturbed at a time when they are wishing to lay.

Obviously, grossly mis-shapen eggs will probably not hatch. Slightly lop-sided eggs are frequently produced by over-fat hens and these will hatch. Contrary to the belief of some, pale blue eggs, which are usually thin shelled, are capable of hatching.

G. EGG STORAGE.

Eggs are best rested for 24 hours between the time they are laid and when they are set. Storage for longer than 24 hours will lead to a gradual and then a rapid loss in hatchability. The rate of deterioration will depend upon storage conditions. Under ideal conditions, the difference in hatchability between a two day and a seven day stored egg may be in the region of 4 per cent. In poor conditions, this may rise to 10 - 15 per cent. This is one of the reasons why many large pheasant hatcheries set eggs twice a week. An ideal egg store provides clean, still air conditions with a temperature of between 55 and 65°F and a relative humidity of 75 per cent. Very few pheasant egg producers have good egg stores.

Eggs should only be placed in the store when they are dry. If they are only to be stored for a maximum of one week, they should be placed blunt end up and will derive no benefit from being turned. At the beginning of the season, when egg production levels are low and building up, it is quite acceptable to bank a fortnight's supply of eggs before setting. Eggs laid in the first week of this fortnight will benefit from being turned daily. If this cannot be done, store them pointed end up. One must accept that eggs stored for longer than a week will not have good hatchabilty and, in any event, early season eggs are unlikely to be as fertile as later eggs. However, some chicks will be obtained.

It is not essential, but is nonetheless desirable, to remove eggs from store and allow them to warm up gradually before setting them. This should not be done too rapidly because condensation may form on the shells, increasing the likelihood of bacterial infection.

H. PROBLEMS AND THEIR PREVENTION.

i) Internal Parasites.

Protozoal parasites such as coccidia, hexamita and trichomonas are rarely of consequence in laying pens. Most birds of breeding age should be immune to ill effects although they may well be carriers of these infective agents.

The three helminth, or worm, parasites that may sometimes be important are Syngamus (gapeworm), Heterakis (caecal worm) and several species of Capillaria (thread worms). If gapeworms are present, it should be obvious by the snicking and gaping of the birds. Pheasants often appear to lose their immunity to Syngamus at point-of-lay, probably due to hormonal changes.

Heterakis and Capillaria problems are not clinically obvious, but should be investigated in cases of poor performance. It is generally agreed that the presence of very large numbers of Heterakis (several hundred per bird) is harmful. It is much more debatable as to whether smaller numbers of up to 50 per bird are of any consequence. It is these lower numbers that are normally encountered and, while they may affect wild pheasants, struggling to maintain a reasonable level of nutrition, they should not affect birds in laying pens fed adequate quantities of properly formulated breeder diet. Heterakis are small worms which are nevertheless visible to the naked eye when the caeca or blind guts of infected birds are examined. Heterakis are sometimes involved in the transmission of the protozoan parasite, *Histomonas meleagridis*, which causes the disease known as blackhead. Blackhead is, however, rare in pheasants and, in any event, many breeder diets contain anti-blackhead drugs.

Capillaria species are very thin worms, often well embedded in the gut wall, and cannot be readily seen by the naked eye. Several species occur in pheasants and the crop, intestines and caeca may all be infected. The species affecting the crop probably cause the most damage to the birds. In order to establish the presence of capillaria infection, veterinary diagnosis is necessary. It is not common to encounter heavy infections of Capillaria species in laying pens and it would be unusual, therefore, were they to cause problems.

In summary, the gapeworm is usually the only significant internal parasite of the laying pen and, when its presence is detected in appreciable numbers of birds, treatment is justified. It is not sensible to treat routinely in the absence of signs any more than it is to take an aspirin to prevent a headache. However, it is reasonable to treat pheasants in winter quarters if even a few of them are gaping before moving them to the laying pen. This will prevent the contamination of the laying pen with parasite eggs although these may already be present from the previous season. Under other circumstances, do not be in too much of a hurry to treat the flock. A few gapeworms in the trachea will do no great damage and, if one waits until significant numbers of birds are infected, it is less likely that treatment will have to be repeated. There are only two non prescription drugs available for the treatment of gapes in breeding stock - fenbendazole (Wormex) and flubenol (Flubenvet). The former is added on to feed and the latter has to be incorporated into feed by a compounder.

ii) External Parasites.

Lice are not uncommon. They chew feathers and other surface debris on the skin, depending upon their species, but, unless present in very large numbers, do little harm. When large numbers are present on a bird, it is usually an indication that the bird is sick for other reasons and is not preening itself properly.

Mites, which are much smaller than lice, but which can just be seen with the naked eye, may cause severe damage when large numbers are present. Some species suck blood and they cause much irritation to the birds. Often the first sign of mites is irritation of egg collectors when the mites get onto their arms. The red mite is only to be found on the birds at night and spends the day in cracks in woodwork round nests and perches. Other species spend their whole lives actually on the birds.

Mite infections are not common in laying pens. When they do occur, they are not easy to deal with and veterinary advice should be sought.

iii) Bacterial and viral diseases.

Marble spleen disease is caused by a virus - the same as that which causes haemorrhagic enteritis in turkeys. This virus is very widespread in pheasants although it was rare ten years ago. Many pheasants are immune to it and many others become infected and suffer no adverse consequences. However, when pheasants receive a very heavy challenge or are stressed, having just been infected for the first time, they are apt to die extremely suddenly. Post mortem examination will reveal enlarged spleens and very darkened, fluid filled lungs. The birds are usually in very good condition and die of suffocation due to the fluid on the lungs. There is often a coincident history of the birds having been frightened at night by, for example, a flashing torch or a fox prowling the periphery of a pen. It is probable that, without such a stress, the birds would have survived infection with no untoward effects. Death from marble spleen disease is rare before eight weeks of age, but can occur at any time thereafter. Rates of mortality can vary from very low to over 50%. Where the rate is high, it is probable that a large proportion of the pheasants are experiencing the virus for the first time which can occur when poults or breeding stock from different sources are mixed together. Within a few years, given the already rapid spread of the virus among pheasants, infection will probably become universal. In such circumstances, all breeding stock will pass immunity to their chicks which, in turn, will become infected while protected by maternal immunity and themselves develop active immunity. When this occurs, it is likely that the disease will more or less disappear.

Other bacterial and viral diseases are rare in laying pens. A few birds usually die of egg peritonitis during the course of a season and others of impacted oviducts. Mortality from these causes will rarely exceed 1% and nothing can be done to treat affected birds. Often such birds are ill for quite long periods and go thin. One ought to cull them rather than to leave them to a lingering death. Another cause of chronic illness is avian tuberculosis, particularly in second season breeders. This has been referred to earlier under 'Selection and Acquisition of Stock'. Affected birds cannot be treated and should be culled. In very exceptional circumstances, losses may reach 15%.

Acute outbreaks of bacterial or viral disease, such as erysipelas or fowl pest, can occur and can cause heavy mortality and severe drops in egg production. When they do occur, which is fortunately uncommon, appropriate veterinary advice should be sought as should be the case whenever significant numbers of unexplained deaths occur. In all cases of death, bodies should be disposed of as quickly as possible.

iv) Mycoplasma infections.

Such infections have been discussed in the section entitled 'Selection and Acquisition of Stock'. The disease is chronic and usually characterised by swelling round the eyes. It may lead to somewhat lowered egg production, but will seldom lead to mortality in breeding stock. However, affected birds should be culled because a small proportion of eggs laid by infected hens will themselves be infected. When such eggs hatch, the disease can quickly be spread to other chicks in the hatcher or in the brooder sheds. Flocks often contain carriers which themselves show no signs of disease so one should not assume that the problem has been solved by the culling of visibly affected birds. Veterinary advice should be sought in these cases and the veterinary surgeon may well consider it appropriate to treat the flock intermittently with an appropriate antibiotic to reduce the chances of egg transmission.

v) Treading injuries.

Treading injuries tend to be by far and away the greatest cause of loss in hen pheasants in communal laying pens. Severe injuries can be inflicted on the backs and flanks by the spurs and inner toenails of the cocks. If the injury is not too severe, the hen may recover if removed from the pen. If left there, she will probably be cannibalised. Severely wounded birds should be culled at once. Losses through treading may, on occasions, reach 20%.

There are various ways of attempting to prevent the problem. The most usual is to blunt the spurs and all of, or just the inner, toenails of the cocks before placing them in the pen. This operation is most often carried out with the hot blade of a debeaking machine. One should endeavour not to chop the nails or spurs square or this will leave damaging sharp edges. However, even when the cocks have been dealt with in this way, treading injuries can still occur. Cock pheasants are dominant over hen pheasants during the breeding season. Hen pheasants have their own pecking order. The more dominant hens are trodden far less frequently than the hens at the bottom of the pecking order. The dominant hen is thus trodden when she wants to be and, therefore, doesn't struggle and is rarely injured. The unfortunate hen at the bottom of the pecking order is pushed from one part of a communal pen to another and is repeatedly raped by a succession of cocks. As she does not appreciate such lavish attentions, she struggles and is subject to injury. The provision of adequate space, 'baffles', cover and perches will give such hens some respite.

Cannibalism following treading, which usually kills more hens than the treading injuries themselves, is greatly reduced when both hens and cocks are fitted with 'spex'. The backs of slightly injured hens can be protected by the fitting of saddles which are held in place by tapes round the wings. These can be home made, but are now also available commercially. Birds thus protected may be left in the pen to lay.

vi) Vent Pecking.

Probably second in importance to treading injuries as a cause of loss of laying hen pheasants is vent pecking. When a bird lays an egg, it everts its oviduct through its cloaca and vent, exposing a deep purple, velvety tissue. This is highly attractive to other birds which, given the chance, will peck at it causing bruising and even bleeding. Such damage interferes with the prompt retraction of the oviduct and it can thus remain exposed to further injury. At its worst, vent pecking can lead to the entire oviduct and even the intestines being eaten out of the living bird. It is a vice which is habit-forming and spreads among the flock. Death may be much more long drawn out than in the example described above. Sometimes, the cloaca becomes blocked and bruised and faeces impact internally while white urates dribble down between the bird's legs. Such birds lose condition and should be culled. Sometimes, just the vent is damaged, but the bird remains otherwise well. However, even this degree of damage will seriously interfere with successful mating and thus result in lowered fertility.

Prolapsing of the oviduct can predispose to vent pecking and help to get the vice established. Prolapses occur most often at the start of lay and are more likely in fat hens. By the time a prolapsed bird is found, its vent region may have been so seriously cannibalised that the original prolapse is no longer discernible.

There are two important ways to combat vent pecking. If the hen is allowed seclusion in which to lay, the problem is greatly reduced. A sufficient number of well designed nests is thus of the utmost importance. The second way of overcoming the vice is to fit both cocks and hens with 'spex'. In a trial involving several pens of laying birds in which the birds in half of the pens were fitted with 'spex' and half were not, the mortality through vent pecking was 3% in the latter group and only 0.25% in the former. Furthermore, fertility was 5% better in the eggs produced by the groups in 'spex'.

vii) Poor Egg Production.

As has been stated previously, egg production rarely reaches its true potential. Poor performance is often blamed on the weather, which may sometimes be a valid argument, or the diet which is very rarely so. However, failure to provide adequate feeders, drinkers and baffles in the pen can all lead to suboptimal feed or water intake and generally unhappy, stressed hens. It is in this area that most of the problems lie.

Not to be forgotten, however, are egg predation by corvids from open topped pens and egg eating by the pheasants themselves. In the latter case, partly eaten eggs or egg shells are often found, but many eggs disappear completely. Filling eggs with mustard, Friar's Balsam or other supposedly unpleasant tasting substances is seldom helpful. Pheasants, in fact, often appear to relish mustard almost as much as eggs! Some keepers shoot pheasants that they find egg eating. However, since many pheasants will come to peck at an egg once it has been broken open, this action is almost certainly both costly and pointless. The answer primarily lies in the provision of adequate nests, preferably bedded in deep, loose straw, and frequent egg collection. The fitting of 'spex' also reduces the incidence of egg eating.

One should never overlook the obvious. Drops in egg production are occasionally caused by escapes of hens from pens through holes in the perimeter wire or loss of brails. Loss of hens through predation is also obvious, but is scarcely likely to be overlooked.

viii) Small Egg Size.

Early season eggs tend to be smaller and a higher proportion are rejected as being too small to set than those laid later in the season. If the problem is severe and persistent, suspect a lack of protein or essential fatty acid in the diet or a disease problem which interferes with nutrient uptake.

Problems relating to fertility and hatchability, though often arising on the laying farm, will be discussed in the next chapter which deals with incubation.

I. STATISTICAL GUIDELINES.

Approximate breeder feed consumption per 100 hens and appropriate cocks from two weeks pre-lay till mid June	: 800 kg
Acceptable numbers of eggs/hen from start of lay to mid June	: 40 plus
Percentage of unsettable eggs	: Less than 8%
Acceptable laying pen mortality	: Less than 5%

J. RECORD KEEPING

The keeping of records is an important discipline. Records are an invaluable starting point in the investigation of problems. The following information should be recorded:

   a) Numbers of hens put into laying pens.
   b) Numbers of cocks put into laying pens.
   c) Mortality and, if possible, cause.
   d) Weekly feed consumption.
   e) Eggs collected/day.
   f) Eggs set or sold.

CHAPTER 2
INCUBATION

A. HATCHERY DESIGN.

When establishing a hatchery, considerable care should be taken in its design. The bigger its size, the more important this is. Thought should be given particularly to the establishment of hygienic and labour saving working practices. Skimping on capital expenditure is extremely tempting, given the fact that pheasant hatcheries are normally idle for three quarters of the year. Interest charges on capital and depreciation often together amount to over 50 per cent of the cost of hatching eggs.

In a well designed hatchery, clean eggs should enter in one direction and chicks exit from another. There should be a minimum of cross traffic. The most important consideration is that, in all but the smallest hatcheries, the setting machines should be in a separate room from the hatching machines and that the two areas should have independent ventilation systems. The hatching area is a dirty area and dust and bacteria released at hatching must not be allowed back into the setting side of the hatchery. In larger hatcheries, which hatch twice a week, there should be two hatching rooms.

If eggs are to be stored at the hatchery rather than at the supply farms for longer than overnight, a proper egg store with temperature and humidity control is highly desirable. Larger hatcheries may fumigate all incoming eggs on arrival, necessitating a proper fumigation cabinet with appropriate extractor fans so that the formaldehyde gas can be voided to the outside of the building.

It is important that the chick boxes and liners or wood wool, etc., be stored in hygienic conditions away from wild birds, rodents or dust.

There should be an exit from the hatching side of the hatchery to allow contaminated hatching equipment to be taken out and cleaned, preferably with a pressure washer or steam cleaner. Unless all organic matter is completely removed, disinfection is unlikely to be successful.

B. SETTING.

Pheasant eggs are hatched in a huge array of equipment. It is beyond the scope of this text to cover all the different types available and, therefore, only general principles will be discussed.

Between the time they are set and that at which they are transferred to hatching machines, eggs have the following requirements:

a) The maintenance of a temperature of between 99.5 and 99.9°F at the centre of the egg.

(If one is setting in still air machines, temperature is sometimes read at the level of the top of the egg at which point it might need to be 103°F due to the temperature gradient from top to bottom). Even temperature differences of as little as 0.2°F can change average hatch time by several hours.

Most pheasant machines are multi-stage setters, implying that one third of the machine's capacity only is loaded each week and that, at full loading, the machine contains three age groups of eggs. It is extremely important to set trays in these machines in the correct sequence so that eggs of different ages are evenly intermingled throughout the machine. When they are first set, eggs require considerable heat input. At later stages of incubation, they produce a lot of their own heat. If these older eggs are concentrated into one area of the machine, the heat they produce will not be adequately dissipated and local hotspots can occur, leading to premature hatching of some eggs and poor overall hatchability.

It is possible to acquire single stage setters which can be filled completely at one time. These machines, however, are more sophisticated and expensive in that they require to have cooling systems as well as heating.

b) Air movement through the machine.

This is essential to supply oxygen and remove carbon dioxide and water. Movement through the machine is normally achieved by paddles or fans except in the small, still air machines which rely on convection. It is no use moving air through the machine if the air in the setter room itself is not being regularly changed. Otherwise, the same stale air - or a proportion of it - is being constantly recirculated. The room itself, therefore, requires a good ventilation system. It is generally considered better to blow air into the setter room rather than to suck it out. This should not be allowed to cause direct draughts to hit the machines themselves. If air is pumped in, it should find its own way out, preferably through vents over the machines.

The rate of air movement through the machine is important because of various interactions. If it is too low, heat dissipation in the machine will be more difficult and may lead to unacceptable temperature variations within. It can also influence humidity requirements. If air movement through the setter is low, insufficient water will be lost and, unless the machine is run at lower humidity, the air spaces of the eggs will be too small the allow satisfactory hatching. Conversely, if air flow is too high, more water must be added to the machine to prevent air spaces from becoming too big.

c) Humidity.

This must be controlled. During its approximately three week period in the setter, an egg must lose in the region of 13% of its weight. This allows an air space of proper size to develop at the blunt end of the egg. The rate at which this weight loss occurs is governed by both humidity and rate of air movement. In some machines, the humidity is automatically adjusted by water sprays while, in others, it is determined either by the presence or absence of water in the water pans or the exposed surface area of the water pans. The automatic system can give problems if water jets block and it is particularly important, in hard water areas, that proper maintenance is carried out.

It is not imperative for weight loss to be constant every day. Therefore, when using manually filled pans, it is sufficient to keep them full for some days of the week and empty on others. The regime may have to be varied according to the dryness or otherwise of the air coming in to the setter room. Regular weighing of trays of eggs is the best way of monitoring one's regime. Some indication, albeit approximate, can also be gained by candling batches of eggs and assessing the sizes of air spaces. If too much weight has been lost too early, correction can be made by adding more water during later stages of incubation and vice versa. However, this can lead to complications given that most setters have eggs of three different ages within them. If the machine is to be run dry for some of the week and wet for the rest, avoid the dry period coinciding with the day of setting and the following day.

d) Turning.

In the first stages of incubation, it is essential that eggs be turned, at the very least, once a day. Three to five times a day is better. Most larger machines turn eggs every hour. This is far more than necessary but does no harm. After 14 days of incubation, turning is no longer required, but, again, a further week of turning is not harmful.

If eggs are not turned in the first week, the developing embryos stick to the shell membranes and hardly any of them hatch. Failure to turn in the second week is not quite so damaging. It should be a routine part of hatchery management to check that automatic turning gear is operational.

e) Hygiene.

It is possibly incorrect to list this as a requirement of the egg. However, only cleaned eggs should be set. Whether one decides to disinfect the setters on a routine basis is debatable. It is possible to fumigate the setters weekly. If this is done correctly, it may do some good. However, if done at the wrong times of the week in relation to setting, it can be very damaging to hatchability. It is probably better to avoid fumigation and, if one wishes to do anything to the setters during the season, to fog them with an iodine based disinfectant. Whether or not one opts to do this, one should completely clean and disinfect the setter and setting trays both at the start and the end of the season.

Occasionally, eggs explode in the setters. Resulting gross contamination should be washed off surfaces with a disinfectant solution which will not corrode metal. Operators should always wash their hands before handling hatching eggs.

C. HATCHING.

It is generally assumed that, unlike chicken eggs, pheasant eggs hatch better in machines in which there is no fierce air movement. Most pheasant hatchers, therefore, rely on convection or a relatively slow fan movement to provide air exchange. During hatching, oxygen requirements/egg are maximised, but it is, nevertheless, desirable to allow a build-up in the concentration of carbon dioxide which acts as a respiratory stimulant. However, an excess of carbon dioxide, due to inadequate air exchange immediately after full air breathing has started, can lead to rather drowsy chicks. These usually quickly recover when removed from the machines and boxed.

It is important not to allow too high a temperature to develop in the hatcher room, particularly when using still air hatchers. Given that air is mainly moved by convection which is driven by temperature difference between the room and the machine, a high room temperature reduces air supply to the hatcher. An optimum room temperature will normally be in the region of 70°F. Fresh air should be allowed into the room at low level and direct draughts on to the machines should be avoided. Air can be extracted over the machines, but, if a fan is used to do this, it must not be too fierce.

It is normal to transfer eggs into the hatchers on approximately the 21st day of incubation at which stage none of the eggs should have pipped. The majority of eggs pip on the 23rd day and hatch some 20 to 30 hours later. The humidity from pipping to hatching should be high to prevent the shell membranes from becoming rubbery and impeding the hatching process. After hatching, the humidity can be lowered. With some machines, this is impossible. However, even in these, the damp chicks quickly fluff up their down.

Many egg producers send their eggs to large hatcheries which either buy or custom hatch the eggs. There is obviously the opportunity for significant disease spread and, all other things being equal, the larger the hatchery the greater the risk. Such hatcheries tend to use large, modern fan-assisted hatchers. These are designed to be easy to clean and it is, therefore, possible to minimise the risks of contaminating one hatch by infections persisting from the previous one. However, if eggs from several sources are placed in a single, moving air hatcher, disease in one infected batch is likely to be spread to the other batches in the machine at the same time. As long as hygiene is good all the way along the line, the weight of infection may well be so low as to cause little, if any, significant disease. It is sensible for custom hatchers to keep records so that they know whose chicks go to which customers. In this way, if disease does occur, it may prove possible to define the source quickly.

Disease is, perhaps, less likely to spread from tray to tray of a still air hatcher although such machines tend to be more difficult to clean.

D. TAKING OFF THE HATCH.

It is general practice to open the hatchers and box the chicks on the morning of the 25th day, the chicks having hatched the afternoon or evening before. If the chicks have not fluffed up and the hatch looks incomplete, delay is desirable. This would tend to indicate that the setters have been maintained at too low a temperature. Sometimes, a few chicks appear on hatching trays on the 23rd or even the 22nd day. This is often a sign that the hatch from that tray will be long drawn out and give poor results. While this may, on occasions, be due to hotspots in the setting machine, it is more likely to be due to eggs having been heated up in nests before collection or having been stored at temperatures which are too high. This will accelerate the hatching of some of the eggs, but weaken others so that they either die in early incubation or develop normally, pip, but do not manage to hatch spontaneously. It is not a good idea to open the hatcher early to remove the few chicks that may have hatched prematurely as it is likely to spoil the chances of the majority of eggs not yet hatched.

As one boxes the chicks, those that are deformed should be killed. If the chicks are to be sold, small ones and those with improperly healed navels should not be boxed. It is normal to pack 2% extra. Once all the fluffed up chicks on a particular tray have been boxed, one will be left with a varying proportion of unhatched eggs and, maybe, a few damp chicks. Of the eggs, some may be pipped. The proportion will depend upon whether or not one has had a good, clean hatch. If one has, there will be few pipped eggs on the tray. However, there may well be many pipped, but unhatched eggs. One is then faced with a decision whether to return them to the hatcher or to break them open. Better overall results are usually achieved by helping the chicks out of their shells there and then. If the quality of the hatch has been good, the quality of the few 'help outs' will be poor. However, if the hatch percentage has been low, there will be many pipped eggs and many of the 'help outs' will develop into totally normal poults. When breaking out these chicks, those whose yolks have not been withdrawn into the body cavity should be immediately killed. Those which are very dry and are stuck to the egg shell membranes will probably have to be culled subsequently as they are likely to have 'wry necks' or crooked feet. Most broken open eggs at this stage, however, will contain moist chicks and there may be a minor amount of blood loss associated with the breaking out procedure. This is usually of little consequence. In cases where most of the pipped eggs first broken open contain chicks which still have yolks outside the bodies, it is best to replace the hatching tray in the machine for several hours before breaking open the rest of the eggs. The helped out chicks should be put back into a hatcher to dry out with other moist chicks. Several trays-worth of them can be grouped on to a single tray. The whole procedure may sound unhygienic and, undoubtedly, to some extent, it is. However, experience over many years suggests that the breaking out of pipped eggs very rarely causes significant health problems later. Once a hatching tray has been processed in this way, it should be removed from the hatchery. Unhatched eggs should be put into a bag and the tray soaked in water prior to pressure washing. This is a dusty process and should not take place within the hatchery proper. The surface upon which the tray had been placed for dealing with the chicks should be well disinfected and the operator's hands should also be washed in disinfectant before proceeding to the next tray.

When one has boxed all the fluffed-up chicks, one will be left with a few trays containing chicks which had previously hatched, but not been dry enough to box and chicks which had been 'helped out'. These should be left for four to six hours to dry off. They can then be graded. Some will have already died, some will be on their sides unable to stand, some will have stiff or crooked necks while others will have spraddled legs or clubbed feet with inward turning toes. All such chicks should be destroyed at once. The remainder, which may constitute between 50 and 75 per cent of the total, should be boxed and put under brooders as soon as possible. It is not ethical to sell such chicks since they sustain a somewhat higher mortality than first quality chicks in the first week of brooding and require a rather higher rate of culling later due to deformities missed on the hatching trays.

After the hatch has been completed, the hatcher room and equipment should be cleaned as soon as possible. A vacuum cleaner can be used to reduce dust during cleaning. Having cleaned all equipment, it should be disinfected with an approved proprietary compound. Ensure that the disinfectant used does not corrode metal or the life of the equipment will be greatly reduced. Attempt to have the cleaning programme completed by the end of hatch day so the room will receive a break of two and a half days before transferring more eggs into it.

It should go without saying that cardboard chick boxes should be new and never re-used. Plastic boxes for internal use should be scrupulously cleaned and disinfected before being used again.

E. HATCH ANALYSIS.

The term hatchability can be defined in several ways. This can lead to confusion when results are compared or problems investigated. Suppose that 1000 eggs are set and that 630 first quality chicks are boxed. Suppose that a further 150 chicks are 'helped out' and that 60 per cent of them are subsequently brooded, the rest being killed on the hatching tray. What is the hatchability? One could say that the hatch of first quality chicks is 63 per cent and ignore the 'help outs'. Equally, one could count the 630 chicks of the first quality and the 90 surviving 'help outs', giving a hatch result of 72 per cent. However, the 'help outs' will not rear as well as the first quality chicks. Possibly, the fairest assessment of hatchability, therefore, is to count only 90 per cent of the surviving 'help outs'. This gives a result 630 + 81 out of 1000 - 71.1%. This is only a fair assessment if one grades the 'help outs' severely on the hatching trays.

Some people candle their eggs and reject the clear eggs. If they express their hatchability in terms of the eggs that were not clear, they will obviously get a higher figure that when expressing in terms of all eggs set. This higher figure is often called 'the hatch of fertiles'. However, this presupposes that all the clear eggs are infertile and this is not the case as many of them may have died at an early stage of incubation.

On occasions, it is worth breaking open some of those eggs that do not hatch to attempt to discover why. If there is a hatchability problem, it is particularly important to do this or to have somebody else do it. If one returns to the hypothetical hatch discussed previously, in which 630 first quality chicks hatched and there were 150 'help outs', there would remain 220 unhatched eggs. It would be tedious to break them all open. Suppose one were to break open 100, one could expect to find the following:

36 infertile eggs

20 'early dead germs' with no evidence of an embryo in the egg

6 with small dead embryos - 'middle deads'

20 fully developed dead embryos which had not entered the air space

18 fully developed dead chicks with beaks in the air space

This makes a total of 100 eggs - all that were broken open. Each figure on the list should next be multiplied by a factor of 2.2 because there were 220 unhatched eggs. 36 multiplied by 2.2 equals 79 to the nearest whole number. There were thus 79 infertile eggs in the total of 1000 set. The fertility was, therefore, 1000 − 79 = 92.1%. This is the true fertility figure. The fertility based on the assumption that all incubator clears are infertile would give a different result which would include all infertiles and all 'early dead germs'. This would work out as follows: (36 + 20) multiplied by 2.2 equals 123. Fertility based on clears is, therefore, 1000 − 123 = 87.7%, the difference being made up of 4.4% of early dead germs. Since the causes of infertility and early death of the embryo are different, it is important to distinguish between the two.

When breaking open an infertile egg, its contents appear relatively normal although there may be some pale discolouration to the surface of the yolk if the egg has been incubated for 25 days. If one pours out the contents, the white will be clear and 'runny'. In an early dead germ, the yolk may or may not look abnormal, but will tend to have a more discoloured surface than the infertile egg. If the egg died after the fourth day, evidence of blood will be detected. When tipping out the contents, some of the white will be more solid and gelatinous than that of an incubated, infertile egg. Although not of immediate relevance to this discussion, it is worth noting that one can distinguish between fertile and infertile eggs by breaking them open before incubating them. This can be done on unsettable eggs as a rough and ready fertility check of, for example, a cock in a single harem pen. If one carefully opens the top of an egg at the blunt end and looks at the yolk, one will see a solid, whitish spot which represents the unfertilised ovum. If, however, the white spot is closely surrounded by a whitish ring, the egg is fertile. The ring is the extremity of the developing blastodisc or early embryo. An egg spends about 25 hours in a hen before it is laid. It is fertilised at the beginning of this period and continues to develop thereafter so that, by the time it is laid, there are many cells in the blastodisc and thus one can actually see it with the naked eye.

F. EXPECTED HATCHING RESULTS.

If one expresses hatchability on the basis of all first quality chicks and 90 per cent of non culled 'help outs', an average figure for the whole season of around 70 per cent would be typical. It would be lower early and late in the season and higher at peak. The 'help out' figure should be a minor contributor to the total percentage (less than 10%).

The range in hatchability, however, is very great. Suppose a hatchery takes in eggs from six suppliers, it will probably have one or two who produce consistently better results than the average and whose eggs hatch at 75 to 80%. Often, feed can be eliminated as a source of this superior performance as it is often the same as that used by suppliers of substandard eggs. One tends to notice that the hatchability of eggs from the best suppliers varies less from week to week within a single hatching season than those from one's more average suppliers. The latter may have one or two excellent hatches, interspersed with others that are up to 10% below expectation. This would imply that most batches of eggs never achieve their full hatching potential and that their performance has been impaired even before they reach the hatchery. There is also much variation in the potentials of different hatcheries.

G. HATCHING PROBLEMS.

When a group of eggs hatches badly, it is necessary to break open unhatched eggs and assess at what stage the problems occurred.

If the poor hatch is due to an increased percentage of clear eggs, the source of the problem is almost certainly not to be found at the hatchery. Clear eggs are either infertile or contain early dead germs.

True fertility in the best flocks averages 95% in mid season. If fertility is below 90%, there are probably behaviour problems involving dominance and aggression in communal pens. Vent pecking may also be a cause. Usually, when fertility is substandard, there is a simultaneous increase in the number of early dead germs. This is due to the fact that, in cases of poor fertility, some eggs are fertilised with stale or substandard sperm. These may start development, but die soon thereafter. A good batch of eggs will contain no more than 5% infertiles and 3% early dead, a total of 8% clear eggs. One often encounters batches of eggs with double and sometimes even treble this percentage.

An increase in the numbers of early dead germs is not always associated with fertility problems. Other faults which cause early deads are as follows:

a) Eggs chilled on nests overnight in early season;

b) Eggs overheated on nests due to infrequency of collection and overuse of nests by hens;

c) Eggs overheated or roughly handled during washing;

d) Poor temperature and humidity control in the egg store;

e) Setting stale eggs;

f) Bacterial infection of eggs.

Most of the above faults may also cause an increase in late deaths. However, where problems are confined to an increase in late deaths with no coincident rise in the numbers of clears, the problem is most likely to have developed in the setting incubators. One should be looking to hatch 85% of non clear eggs. If, in a good batch of eggs, 8% are clear, the hatch of all eggs set will thus be 92 x 85 = 78.2%. Commonest setter faults arise through lack of adequate air supply or poor control over humidity. By breaking open eggs and examining late deads, it is possible to get an idea as to air space size. If it is small and there is still a lot of fluid in the eggs, there has probably been too much humidity during setting. The chicks that do hatch will tend to be heavier and somewhat floppy. One sees the opposite situation in cases of inadequate humidity. Overheating during setting will lead to premature hatching and under-heating to late hatching.

Complete failure of turning gear in the early stages of setting can lead to a huge increase of late dead germs. An expected hatch of 70% may drop to 15%. If the failure occurs in the second week of incubation, however, the hatch may only be 5% down.

It is rare to see large numbers of deaths of embryos in the middle stages of incubation. However, when this does occur, one should suspect a nutritional deficiency in the breeder diet.

Despite enormous attention paid to hygiene in relation to eggs and incubation, very few hatching problems are due to infections with bacteria or moulds. This is not to say that occasional eggs are not killed by infection - they are. The real importance of hygiene is to minimise bacterial and other infections in the chicks that hatch and then spread their infections to healthy chicks hatching in the same environment either at the same time or in subsequent hatches.

CHAPTER 3
REARING

A. HOUSING.

i) General Principles

Those who are familiar with modern intensive poultry rearing systems are often fairly scathing about the methods used to produce pheasants. The latter are vastly more labour intensive and, often, mortality to six weeks in pheasants is much higher than it is in, say, broiler chickens. Many gamekeepers and shoot proprietors, on the other hand, are apt to consider that anything remotely resembling a poultry system of production must automatically be bad and that grass reared poults are inevitably bound to be superior to those produced by any other method. Both viewpoints are exaggerated and neither is correct.

In order to resolve the above prejudices, it is necessary to consider the different aims of broiler chicken and pheasant poult production and also the means available to achieve those aims. A broiler chicken must grow at maximum rate for a minimum amount of food with very few labour inputs. This requires that it is produced in a fully environmentally controlled, windowless building. As soon as it is removed from the building, it is slaughtered. The purpose of a rearing programme for pheasants is to take day-old chicks and turn them into poults ready for release at six weeks of age or shortly thereafter. While this should be achieved as cheaply as possible and with a minimum of mortality, it is essential that the released poults should be hardy and weatherproof. This requires that they should be well feathered, that they should be used to natural day and night conditions, that they should not have received artificial heat for at least ten days before release and that their plumage should have been wetted several times. Clearly, they should also be healthy and in good body condition. It follows, therefore, that it is considerably more taxing to produce a satisfactory pheasant poult than a satisfactory broiler chicken. Despite this extra difficulty, the housing systems available for pheasants are usually very poor because they are empty for most of the year and thus cannot justify high capital expenditure. One must thus compensate for this inferiority with greater levels of stockmanship and labour input. It is the author's view that good quality pheasant poults which are hardy and survive well after release, can be produced from a variety of systems with or without access to grass. It is totally naive to assume that 'grass reared poults must be best' although it is, of course, true that poults produced from a broiler-type system, with no experience of natural lighting and no weaning from artificial heat, are almost certainly going to be unsatisfactory.

ii) Housing Systems and Stocking Densities.

a. Moveable huts, night shelters and grass pens.

This is a very common method of rearing and excellent results can be achieved from it with good management(*fig 3*). Chicks can be stocked in the brooder hut at a density of 60 - 70/square metre (6-7/square foot). Because of this high stocking density, the chicks should be allowed access via a draught proof pophole to the 'night shelter' after four to five days. The night shelter should provide a floor area double that of the hut. The grass run, leading off the night shelter, should provide a minimum area of 0.2 square metres per bird (2sq.ft/bird).

An alternative to the moveable wooden brooder hut is a purpose built, portable, circular metal brooder heated by paraffin. This tends to provide even less space per chick if used at manufacturers' recommendations. It is, therefore, desirable to provide a weather and draught-proof area immediately outside its popholes.

b. Permanent brooder house with outside grass runs.

The house can either have a central passage with brooder pens leading onto grass runs on either side, or alternatively, the passage may be along one side of the house with a single row of pens and runs (*fig 4*). In cases where the grass runs are fully exposed to the elements, one should stock the brooder pens at no more than 20 chicks/sq. metre (2 chicks/sq.ft.) and provide at least double that area as grass run. However, if the first part of the outside run is designed as a night shelter, one can stock the brooder pen itself at 40 chicks per square metre (4/sq.ft.) and provide an equivalent area as night shelter. The exposed grass run should provide a minimum of 0.1 sq. metres per bird (one square foot).

c. Permanent brooder house with fully covered runs.

The brooder house and internal pens in this system are as described above. However, the popholes or external doors to each pen lead out to a fully roofed, but naturally lit and ventilated area (*fig 5*). The overall maximum stocking rate should be 15 birds/sq. metre (2/3 of one sq. ft. per bird). One third of this total area should be brooder pen and two thirds covered outside run. In other words, one would initially stock the brooder pen at no more than 45 birds/sq. metre (4.5/sq.ft.).

d. Brooder house with no runs.

It is difficult to acclimatise poults for release in such accommodation, but

Fig. 3
A typical outdoor rearing unit with brooder hut, shelter pen and grass run.
(Photo courtesy of the Game Conservancy)

Fig. 4
Indoor brooding units with outside shelters and grass runs.

not impossible. It is an expensive system and, therefore, has little to recommend it unless surplus farm buildings are available. Stock at no more than 15 chicks per sq. metre (1.5/sq.ft.). One must be able to provide natural light and very good ventilation at the later stages of rearing in this system.

e. Mixed systems involving movement of chicks from one unit of accommodation to another.

The previous systems described are all designed to take day old chicks and to rear them to the age of release. There can be some advantages in rearing chicks intensively in an environmentally controlled unit for a period and then transferring them to a more extensive system. The move between systems would typically either occur at ten days or three weeks. On the rearing field at the Fordingbridge headquarters of the Game Conservancy, the former approach is adopted. The chicks initially have the advantage of high quality accommodation with controlled lighting which enables them to get off to a good start with a minimum of early losses and feather pecking. They can be stocked very heavily at this stage. At ten days, they are caught up, beak tipped and moved to the type of system described under Section ii)a. They have to be handled again at three weeks to be bitted. By delaying the move to three weeks to coincide with bitting, the disadvantage of double handling is overcome, but more high quality intensive space per chick is needed.

Fig. 5
Covered outside run used in conjunction with an indoor brooder pen. In this type of system, the plumage of the poults should be sprayed with water several times before release in order to weather-proof them. Provided that this is done and the pens are not overstocked, excellent results can be obtained.

These types of mixed systems are probably over sophisticated for those keepers who only rear one or two batches of poults per year. However, for those who hatch their own eggs and are thus faced with a weekly succession of chicks throughout the season, such systems should be given serious consideration. For further discussion of this subject see under Section iv) of this chapter.

iii) Batch Sizes.

The moveable hut, night shelter and grass run system provides only a small area of heated accommodation per bird. The poults will not all readily fit into the brooder hut as they get bigger and will keep warm at night by huddling together in the night shelter. Large group sizes should, therefore, be avoided with this system. It is recommended that no more than 250 chicks should be started in any one hut. In other words, two units of 250 will generally produce better results than one unit of 500 even though the area per bird may be identical. In the latter case, 'pile ups' of chicks are likely, leading to smothering. Even if these do not occur, a large heap of jugging poults results in a lot of feather loss along the backs of the birds. The ones at the centre scramble over their fellows to avoid overheating while those at the periphery move in the opposite direction in order to warm up.

In systems based on brooder pens in houses, with or without covered outside runs, there is normally more protected area per bird and group sizes can, therefore, be bigger. Five hundred is a reasonable group although numbers up to 1000 are sometimes successful. However, the bigger the group, the greater the management skill needed and the greater the potential for problems. If starting from scratch, the author would design pens to accommodate no more than 350 poults.

iv) Seasonal Stocking Rates.

Section ii) dealt with stocking rates for particular batches of birds, not a succession of batches within a given season. One of the reasons why poults are relatively expensive is high capital cost of rearing accommodation and equipment relative to the very short rearing season. If one stocks a rearing unit in the first week of the season, one can restock it seven weeks later, having removed the first batch and cleaned out. This will allow the unit to be used twice. However, most chicks are produced mid-season and a unit of accommodation for such chicks, assuming they remain within the unit for the full six week rearing period, can therefore only be used once per year.

It is possible, to a limited extent, to tighten the overall seasonal stocking rate by shutting poults out of their brooder compartments at four and a half weeks and, having cleaned these out, to restock after five weeks. Poults should not be shut out in this way, however, unless some alternative covered accommodation such as a night shelter is available to them. Equally, the area of a night shelter would need to be larger than recommended in the earlier section to make up for the loss in the area of the brooder section.

REARING

As a general rule, a game farmer can budget on producing no more than one and a half times the number of poults over a season that he has accommodation for at any one time. A gamekeeper, by buying in only early and late season chicks, can obviously rear two batches per season.

Brooder pens within a house or brooder huts can be reasonably cleaned out between batches as can covered outside runs provided that they were well bedded so the poults had not had direct access to the soil. Grass runs, however, should never be used more than once in a single season. This is an important point of hygiene which, unfortunately, is not infrequently ignored to the detriment of bird health. Second batch poults on the same grass are bound to experience a much greater coccidial challenge than those of the first batch. The anti-coccidial drug in the feed is included at a rate which allows some coccidial development so that the poults will be able to acquire an active immunity. It is not included at a level which would prevent disease in the face of a high challenge. It is quite unacceptable to attempt to overcome this problem by giving extra anti-coccidial medication through the drinking water. Coccidiosis is not the only problem likely to be experienced by second batch grass reared poults.

When using a system of moveable huts, night shelters and grass runs, it is quite simple to provide new grass for the second batch of poults. It is more complicated when grass runs extend from a fixed brooder house. The only way of overcoming this problem is to provide two grass runs for every brooder pen, one for each batch. In order to be large enough, each run must be double the length but only half the width of that which would have been the case had the system been designed for a single batch of poults per season. Within the brooder house, therefore, the pens should be rectangular in shape with the long side of the rectangle running the length of the house and the short side from service corridor to outside wall. With any other arrangement, each of the two grass runs per pen would be too narrow and too long.

In the poultry industry, the so called 'all in all out' method of stocking is used as a vital aid in disease control. In essence, this means that all sites are filled over a short period and then emptied more or less together, enabling a thorough clean out of the whole site before restocking. This avoids the danger of having a multiplicity of age groups on the site simultaneously. This policy is not really practical with pheasants and, in any event, is less important because all pheasant rearing sites should be out of commission for a minimum of 32 weeks a year, assuming stocking starts with first hatch chicks and destocking is complete six weeks after the last hatch. However, it is important, when operating a fixed brooder house, to attempt to avoid having a large, common airspace for chicks of widely differing ages. If, for example, one stocks a house over a four week period at the beginning of the season and then restocks as individual pens become empty, the second batches through are likely to be exposed as chicks to a high level of airborne pathogens. These will arise from dust created when mucking out pens of departed first batch poults and also from other first batch poults not yet moved. This is a perfect formula for creating respiratory problems. There are two ways of minimising this type of situation. The first, practised by many keepers, is to avoid rearing of too many different age groups in any one season. If such keepers produce their own eggs, they can sell them and buy back appropriate numbers of chicks in relatively few batches. While this minimises labour input and enhances disease control over the rearing season, one must accept that the mixing of eggs in communal hatcheries does itself increase disease risks from bought-in chicks.

The second way to minimise the problem of spread of disease between age groups is to divide large brooder houses along their length into several discrete air spaces. This requires the construction of solid partitions between groups of pens and across the service corridors. Obviously, there will probably have to be doors in the corridor partitions, but these should be kept shut when not required for immediate passage. Another advantage of such partitions is that they enable different lighting patterns and intensities to be used which are appropriate to the age of the chicks or poults in each air space.

v) Significant Design Factors.

a. Flooring.

Wooden brooder huts often have wooden floors, but sometimes they have no integral floor and stand directly on grass. In either circumstance, two to three inches of bedding should be placed on the ground or floor surface before stocking. One must appreciate that lack of an integral floor, particularly in a hut not placed on perfectly level ground, creates the risk of escapes of chicks or their destruction by rats that may dig in.

Brooder pens in houses and covered outside pens may either have concrete or dirt floors. Concrete is obviously much easier to clean. Floors should be level or the bedding will creep. There is a risk of drowning chicks in concrete floored brooder pens if a pipe becomes detached from an automatic drinker. In a dirt floored pen, a severe flood will undoubtedly make a mess, but most of the water will disappear into the floor, making drowning extremely unlikely. Major floods should only occur overnight as proper supervision by day will correct a problem before it becomes too severe. One can eliminate risk of drowning overnight when using automatic drinkers by cutting off the water supply as it enters the header tank - assuming the use of a reasonably small header tank. This can be done manually on one's last visit to the chicks in any day, or can be arranged automatically with a time-clock operated solenoid valve.

Although dirt floors are not possible to clean properly, they are quite acceptable. They should be generously bedded and, if being restocked for a second time in the season, all the old bedding should be removed and the floors swept. One can then try to disinfect them as a gesture, but one should have little confidence in having achieved much. The real defence for the

REARING

Fig. 6
Internal brooder pen. Heating is provided by an electric hen and an infra red lamp. Note the rounded corner of the pen as an aid to the prevention of smothering. The drinkers stand on or are suspended over wire frames to reduce contamination of the drinking water with shavings and faeces.

Fig. 7
Door between inside brooder unit and outside covered run. A pophole has been cut into the door and draughts are reduced inside by the use of polythene strips. The pophole can be closed as required. Note ramp and rounded off pen corner.

second batch of chicks will be another generous covering of bedding. The bedding should be removed as soon as possible at the end of the season. If one wishes, one can then soak the floor in an appropriate disinfectant made up in a diesel water mix. This should not be done in mid season if one is going to restock quickly. Even at the end of the season, it is of questionable merit. Prolonged exposure to fresh air before the following season is usually sufficient to prevent viral and bacterial disease agents from persisting.

b. Corners.

All right-angled corners in brooder areas or night shelters are potentially dangerous. They act as focal points for heaping with smothered chicks a consequence. They should, therefore, be rounded off with either concrete, hardboard or netting in such a way that chicks cannot get trapped behind (*fig 6*).

c. Popholes and doors.

Care should be given to the construction of doors and popholes (*fig 7*). It is much easier to manage a group of poults if pophole shutters can be operated from both inside and outside. Equally, when using fixed brooder houses with inside pens and outside runs, it is better if there is a door from inside pen to outside run rather than popholes only. The door may itself contain a pophole. Such doors make servicing of poults much easier. Furthermore, in later stages of growth, the doors can be folded back to allow greater airflow into the inside pens provided severe draughts are not created in this way.

Often, there is a height difference between inside and outside, necessitating a ramp for the chicks. Care must be taken in the construction of such ramps. All too often, free standing ramps tend to come away from the outside walls, leaving gaps down which chicks and poults may drop. Significant numbers of birds may die in this way before the problem is noticed.

When designing doors, consider the size of the transport crates. If they have to be moved through a particular door, ensure that it is wide enough to allow their ingress and egress when they are carried level. It is undesirable to have to tip crates on their sides when they are full of poults.

d. Roofs and Ceilings.

Insulation in roofs of brooding areas is highly desirable since it will reduce heating costs and condensation. In low roofed brooder huts, it will also reduce overheating when the sun is hot.

In brooder pens, it is desirable to be able to hang feeders, drinkers, electric lamps, etc., from the ceiling. This requires a sufficiency of load bearing beams into which hooks can be screwed.

e. Perches.

The provision of some sort of perching facilities in outside runs is appreciated and encourages pheasants quickly to learn proper roosting behaviour once they are released. It also reduces bullying during the rearing stage.

f. Handling.
On many estates, the keeper is single handed. The design of his rearing system must readily allow him to catch his chicks and poults on his own. This is obviously of less importance on game farms where several stockmen are usually available for any required handling procedures.

B. HEATING.

i) Gas.
Gas brooders come in a range of sizes suitable for use with batches of 100 chicks upwards. The large brooders, designed for space heating of chicken sheds are not ideal for pheasant chicks. It is necessary to wean pheasants off heat as soon as possible. When trying to do this with very large group sizes, one is likely to lose chicks in smothering incidents. It is, therefore, better to use small gas brooders which provide localised heat in the brooder houses rather than to heat the whole room. This allows chicks to move away from heaters to cooler areas and prepares them gradually to be weaned from heat.

Some of the medium sized brooders are capable of thermostatic control. The smaller ones, generally speaking, are not. However, inclusion of a regulator on the gas line will enable heat output to be controlled manually. One must be aware that, when the regulator is turned low, an unexpected draught may blow out the heater and lead to disaster. All gas brooders - or lines of gas brooders - should be supplied with two gas bottles with automatic switch-over facilities. Always guard against gas running out unexpectedly.

ii) Electricity.
Electric hens and infra-red lamps constitute the main methods of brooding pheasant chicks electrically (*fig 6*). Electric hens come in various sizes to accommodate between 100 and 200 chicks. They are usually on adjustable legs. The chicks must brood directly beneath them to derive any benefit from their very low heat output of between 80 and 160 watts. One must ensure that the legs do not sink into the bedding, trapping chicks beneath. Initially, if the legs are too high, the chicks will heap on top of each other under the brooder and suffocate. It is quite impossible to inspect what is going on beneath an electric hen without lifting it up and disturbing all the chicks.

Despite the above criticisms, the electric hen is an extremely useful and economical brooder - in terms of running costs - so long as it is not relied upon to provide the sole source of heat in a hut or shed. Either the entire air space must be space heated, initially to not less than 75°F, or the hen should be used in conjunction with infra red heaters. The combination of electric hens and infra red lamps works well and is easy to manage. If one starts with just an electric hen and no lamps, the chicks will have to be confined in a ring which tightly surrounds the hen. This leaves inadequate space for feeding and watering. If the ring is larger, many of the chicks are likely to have piled up and chilled to death before learning the benefits of getting under the hen. If there is an infra red lamp hanging over the ring as well as an electric hen within it, then one can deposit the chicks within the ring and not have to worry that they will chill. The bulb in the lamp should initially be a ruby emitter. After a few days, it will pay to substitute this with a dull emitter which emits heat only and not light. This will minimise feather pecking. However, there must be an alternative source of light to enable the chicks to see to feed.

Perfectly satisfactory results can be obtained with infra red lamps alone without electric hens. Over the whole brooding period, electricity costs tend to be somewhat higher when lamps alone are used. Electric hens tend to be extremely expensive for what they are unless one makes one's own. Lamps are cheaper in capital terms, but the ruby emitting bulbs have a short life and often break if accidentally knocked or splashed with water. For this reason, it is dangerous to rely on a single bulb to brood a batch of chicks. It is much safer to double the numbers of chicks in a ring and provide two lamps. The dull emitters are much tougher, but more expensive to purchase. Infra red lamps that emit white light should be avoided as they will precipitate feather pecking within two to three days.

Although infra red bulbs can be obtained in a range of wattages, the most readily available have a capacity of 275 watts. As a rule of thumb, use bulbs of this wattage at the rate of one per 75 chicks at the start of brooding if no electric hens are being used. If a combination is to be used, two bulbs and a small electric hen will suffice for a ring of 250 chicks and two bulbs and a large electric hen for 350. The initial height at which the bulbs are suspended is important. As a guide, suspend at knee height and adjust according to the behaviour of the chicks beneath.

One can use thermostatically controlled canopy electric brooders designed for poultry. These tend to be too powerful for most pheasant uses. There also tend to be fairly fierce swings in temperatures as the thermostats cut in and out. This makes it difficult for the chicks to rest comfortably beneath them.

iii) Paraffin.

Small round, self-contained metal brooders with paraffin heaters are available commercially. These are designed to be used in grass pens and produce satisfactory results although they are hard work to manage.

C. VENTILATION.

One's aim should be the provision of plenty of fresh air with the minimum of draughts. With many of the brooding arrangements currently used for pheasants, it is extremely difficult to achieve this aim. Lack of adequate ventilation leads to the build up of respiratory pathogens and to poor litter conditions while draughts lead to heaping, smothering, poor growth and poor feathering.

Ideally, static brooder sheds should be designed along the same lines as intensive poultry rearing sheds. In other words, they should be fully environmentally controlled - windowless, mechanically ventilated and insulated. Provided that they are run in conjunction with outside grass or covered pens, they are very easy to manage and will produce well feathered, hardy poults with minimal mortality. However, many static brooder sheds are adaptations of old farm buildings and have no form of mechanical ventilation or insulation. Air is often introduced through windows or other such openings and tends not to be able to escape at the high points of the building. It takes a great deal of ingenuity, in such circumstances, to avoid draughts or the circulation of stale air with ammonia build-up. It is not always easy to decide how most economically to improve or adapt old buildings for satisfactory pheasant brooding. It is often worth taking professional advice which is probably best obtained from appropriate livestock housing specialists of the Agricultural Development and Advisory Service of the Ministry of Agriculture. While such people may not know much specifically about pheasants, they will be expert on their ventilation needs which are the same as those for commercial poultry.

Once popholes are opened, a new source of draught can be created unless they are baffled appropriately or strips of clear polythene are hung from them, allowing the passage of chicks but not strong air currents.

It is important, too, that shelter pens should be free of draughts at chick level as well as being reasonably ventilated. It is easy to turn them into mini greenhouses which become excessively hot and humid by day.

D. LIGHTING.

The correct lighting of brooder pens is important for a variety of reasons. Very bright light in the first few days will precipitate wing tip pecking. Miscoloured chicks are usually picked on first, but the vice can rapidly spread and significant numbers of chicks can die over a short period. The aggressors do much more than peck. They actually grasp the wingtips in their beaks and pull for all they are worth. Very dim light in the first few days, however, is likely to increase the chances of 'starve-out' problems. 'Starve-outs' are chicks that never feed and, having used up their yolk reserves of energy, die between the fourth and sixth days after hatching. Ideally, therefore, chicks should be started with moderate light intensity for a fairly long period each twenty four hours and, thereafter, switched to dimmer light conditions until they are bitted. In this way, feather pecking is minimised. Clearly, one does not have the ability to control lighting patterns fully when using brooder huts, shelter pens and grass runs. In more intensive conditions, it is important that there should not be a constant light intensity throughout any twenty four hour period. If, for example, chicks are maintained under ruby emitters for the first three weeks of life, they will be receiving light by night as well as by day. Gas brooders will also provide some light at night. If artificial overhead lights do not supply substantially more light intensity by day than that provided by the heat source alone at night, it is likely that a few chicks will become blind. These will continue to grow normally and may not be detected until caught up for release or even after release. The reason for this blindness appears to be that, with relatively constant light intensity, the iris of the eye does not expand and contract. Fluid is constantly being pumped into the eye and drains out through small pores around the base of the iris. If the iris does not occasionally open and shut, the pores of a small minority of birds may become blocked, leading to increased intra-ocular pressure and loss of sight in the affected eye. Sometimes only one eye is affected, but usually both. Casual inspection will reveal nothing wrong with the eyes although affected eyeballs tend to have bigger circumferences in the vertical plane and flattened corneas.

It is important that poults are exposed to natural light patterns for some weeks before release. Once bits are applied, there should be minimal problems with feather pecking even in bright sunlight until the beaks start growing around the bits. If bits are applied at three weeks, feather pecking may start at any time from 47 days. The problem can be delayed in covered outside runs if hessian is used to keep out direct sunlight. It is important, too, that poults are used to total darkness before release. One should not leave overhead lights on all night in the brooder pen. However, it is often useful to leave them on until after dark when chicks are first given free access to outside runs. In this way, they will tend to be drawn back inside at night, and are less likely to jug outside and chill in very cold conditions. This is not a substitute for shutting chicks in at night which must be done for the first few days after they have been given access to the outside. It is, however, useful for the subsequent few days.

E. BEDDING.

Brooder huts and sheds are probably best bedded with a one to two inch depth of clean, white wood shavings. Shredded paper can be a substitute. Some people use pea shingle, but this is not a satisfactory medium. It is heavy to work with. Faeces and excreta remain on the surface and cake. The shingle cannot readily be cleaned for reuse. When directly under gas heaters or infra red lamps, it can become very hot so that it is difficult for the chicks to find anywhere that they can comfortably sit.

Wood shavings have several disadvantages. They are quite expensive. They are sometimes contaminated with sawdust which can be eaten and impair growth. They are light in weight and are, therefore, constantly getting into drinkers and feeders. To an extent, the latter problem can be reduced by raising drinkers on to wire platforms after a few days and suspending feed hoppers clear of the litter surface as soon as possible. However, since pheasant chicks are initially smaller and are slower growing than chickens, open feed pans and hoppers do need regular clearing of wood shavings for quite some time.

F. FEED AND FEEDING.

i) Nutrient Specifications.

Pheasants have essentially the same nutrient requirements as turkeys during the rearing stage. Maximum or near maximum growth will be achieved with protein levels in the region of 28 per cent for the first four weeks and 24 per cent from four to eight weeks. This presupposes a high to medium dietary energy density and a balanced amino acid profile in the protein component of the ration. Rations for broiler chickens, unlike those of turkeys, would have inadequate protein and excessive energy levels. While it is possible, therefore, to use turkey rations for pheasants from the nutrient point of view this would be neither legal nor desirable because of the differences in drug inclusions (see later).

Many keepers hold the view that modern pheasants are too large to fly well because they are too well fed on pelleted diets with high protein levels. All scientific and experimental evidence contradicts this view. Final body weight is determined by two factors - genetics and levels of fat deposition. Fat is only deposited in significant amounts in the later stages of growth when energy is consumed to excess or when there is too much energy relative to protein in the diet. Since wheat is higher in energy than typical pelleted pheasant diets but is much lower in protein, it is the premature feeding of wheat after release that is likely to produce excessive fatness. Maize is even higher in energy and lower in protein than wheat. Feeding trials with pheasants have demonstrated that the phenomenon of compensatory growth occurs with this species as with most others. In other words, if chicks are grown slowly to start with by the provision of diets with suboptimal protein levels, they will catch up in size later unless they are hand-fed restricted quantities of feed. However, trials have also demonstrated that pheasants grown at maximum rate to six weeks survive better in release pens and thereafter than those grown more slowly. In a series of experiments where growth rates of two strains of pheasants, reared and fed in the same way, were compared, it was consistently found that poults derived from picked-up eggs of wild fen hens were only three quarters of the weight of those derived from heavy game farm stock. Similar differences in size between the strains were present at six months, both in released birds subsequently shot and in stock maintained in holding pens and fed *ad libitum* on pelleted diets.

ii) Physical form of feed.

The starter feed is often offered by compounders in two forms - crumbs and mini-pellets. One must obviously start the chicks on crumbs. Thereafter, they can progress to mini-pellets if desired. However, it is often more convenient to use crumbs right through to four weeks when switching to grower mini-pellets. There is no right or wrong method, but the pros and cons of each should be appreciated.

Switching to mini-pellets at an early age requires that such pellets be very small in diameter and that they are chopped short in length. Some manufacturers' mini-pellets are much bigger than those of others. The early use of mini-pellets probably reduces food wastage. However, those compounders offering their starter feeds in two forms usually charge more than those only offering crumbs. The really important consideration is the time of bitting in relation to the time of switching from crumbs to mini-pellets. If the two events come very close together, there is much more likely to be an increased mortality associated with bitting. For this reason, as well as for the convenience of only having to deal with one type of starter feed, many pheasant rearers use crumbs till four weeks of age.

Whenever one switches from crumb to pellet, the transition should be gradual. For a couple of days, a mixture of crumbs and pellets should be offered.

iii) Feed quantity.

The amount consumed will obviously be somewhat variable, depending mostly upon the degree of feed wastage. Wastage may well be as high as 0.25kg/poult. A certain amount of waste is unavoidable. When starting chicks and offering crumbs in open pans and on cardboard egg trays, a large quantity tends to be scattered as chicks scratch and dust bath in the feed. Also, wood shavings will get into the pans and will have to be removed regularly, resulting in further feed losses. One should attempt to minimise feed wastage, but never by reducing the ability of the chicks to obtain feed easily. For example, one could keep shavings out of the open pans by standing the pans on wire mesh laid on the shavings. Some keepers cover large areas of shavings with hessian to start with. This works well, but the hessian rapidly becomes fouled. Chicks should be trained to use feed hoppers as soon as possible and these should be regularly raised to chick shoulder height. All these things will reduce feed wastage, but will not prevent it entirely. Under no circumstances should changes to feed arrangements be made abruptly. Do not, for example, remove all open pans when introducing hoppers. It is better to remove them one at a time over a period. Equally, do not suddenly relocate all feeding points from, say, the brooder hut to the shelter pen. Such sudden changes will not inconvenience the great majority of chicks. However, a proportion will be less adaptable and may even starve. More likely, they will go hungry for some days, peck at the litter and pick up a large number of coccidial oocysts at a time when they are not eating a diet containing an appropriate anti-coccidial drug. They, themselves, may then die of coccidiosis and, in the process, may contaminate the environment to such an extent that the coccidial challenge to the remaining chicks may be so high as to overcome the defence afforded by the dietary anti-coccidial drug.

When deciding how much feed will be needed, it is a fair rule of thumb to allow 20 to 22 bags of starter per 1000 chicks to four weeks of age. If the average release age is somewhere between six and seven weeks, a further 25 bags of grower feed per 1000 birds will be required. By six and a half weeks of age, a typical pheasant will be eating approximately 45g/day.

iv) Drug Inclusion.

Starter and grower diets will normally contain antibiotic growth promoters. These will have no influence on disease prevention and are just included to improve growth rate and the efficiency of food conversion.

All starting and rearing diets should contain an anti-coccidial drug. There are currently only two such drugs which are permitted for inclusion into pheasant diets without veterinary written directives - so called PML drugs. There is always the probability that, sooner or later, strains of coccidia will develop that are resistant to the anti-coccidial drug currently being used. In the broiler chicken industry, a variety of drugs is used in various permutations of switch or shuttle programmes in an attempt to overcome the consequences of drug resistance. With only two drugs available for pheasants, to one of which there is already quite widespread resistance, no such programmes are possible. The drugs themselves will be discussed in more detail in a later chapter. It should be appreciated that pheasants do not automatically become resistant to coccidiosis as they get older. The fact that they normally are resistant by, say, ten weeks of age is because they have been challenged by coccidia at sublethal levels and have acquired active immunity. If the anti-coccidial drug is too effective, the birds will not develop active immunity and will remain susceptible to challenge at a later age. The anti-coccidial used should, therefore, be used at a level which prevents disease in the face of a normal level of challenge. If disease does occur, it should not necessarily be assumed that the compounder has failed to include the correct level of drug or that there is drug resistance. It is equally possible that poor management has led to excessive coccidial challenge or that another disease has reduced feed intake so that insufficient anti-coccidial drug has been consumed.

Grower diets and starter diets for chicks which spend a proportion of the first month on grass should contain an anti-blackhead drug. The importance of this drug is not so much that it protects against blackhead, to which pheasants are not particularly prone, but that it is an effective way of preventing the consequences of Hexamita and Trichomonas infections which, on occasions, can be devastating.

Diets designed for chickens and turkeys often contain anti-coccidial drugs which are inappropriate for pheasants. Some drugs may be toxic to pheasants and others may be so effective at controlling coccidial infections that no immunity develops. The nutrient specifications of pheasant and turkey diets for birds up to eight weeks of age are, to all intents and purposes, identical. However, because of the differences in drugs used for the control of coccidiosis, turkey diets should not be used for pheasants without veterinary advice having first been sought.

v) Feeders.

However much feed is available to chicks, poor results will be obtained unless there is a sufficiency of easily accessible feeding space. In the early stages, there should be so much choice of feeding points that chicks cannot fail to find feed. In cold weather, chicks may prove reluctant to leave the immediate area of the brooder to venture to the colder regions of the pen to find feed. Equally, it is important, in the first few days, that there is sufficient light to stimulate feeding behaviour.

Later on, approximately one feed hopper per 75 chicks or poults is a reasonable guideline. Ideally, these should be fairly evenly distributed throughout the area available to the birds. This requires that the feeders in grass runs be weatherproof or placed under some sort of roof in the pens. It is not a good idea to place all feeders in the shelter pen just for convenience. Never move all feeders from area to area simultaneously - make all changes gradually.

G. WATER AND DRINKERS.

Chicks and poults will drink between two and three times as much weight of water as they will eat feed. It is vital that chicks can quickly find water when first placed in brooders and equally important not suddenly to move drinkers at later stages so that, for example, water ceases to be available in the brooder hut when first transferring it to the shelter pen. Birds must be given time to learn where the new waterers are before removing old ones. Equally, no sudden changes should be made in the types of drinker in which water is presented. Do not, for example, when using automatic drinkers, remove all the satellite mini-drinkers and replace them with the main drinkers at one time. Water should be available at all times and birds should never be locked away from it overnight.

Chicks are liable to drown during the first few days. Many keepers put stones in the drinkers to prevent this. However, this is labour intensive since the drinkers should be cleaned out frequently to prevent the growth of bacterial slime which takes place rapidly in hot conditions. Even automatic drinkers should be cleaned on at least a daily basis. To prevent drowning, it is easier to put a ring or circle of pipe of suitable diameter into each drinker for the first few days (*fig 8*).

REARING

Fig. 8
A mini automatic drinker containing a ring of pipe to prevent chicks from drowning. This is much more labour saving than the use of pebbles.

Fig. 9
A feather pecked poult. Such poults should not be released until their feathers have regrown. If they are, they are likely to die of exposure.
(Photo courtesy of the late T. Blank)

The use of automatic drinkers is undoubtedly labour saving. However, it does lead to a high probability of flooding sooner or later in the season. Valves and floats may stick and, worse still, pipes may become detached which can occasionally lead to serious drowning incidents in concrete floored brooder pens. It should be ensured that header tanks are not too high as the water pressure will be excessive. To avoid severe flooding the mains supply to the header tank can be cut off overnight. In the case of a detached pipe, the only water to flood will then be that in the tank itself. Some rearers will not use automatic drinkers in the brooder compartments at all and will confine their use to outside runs only. If one is using automatic drinkers it is important, having cleaned them out, to re-examine them carefully soon afterwards because the cleaning process itself is liable to precipitate leaking.

If there is a flood, wet bedding material should be removed as soon as possible and replaced. Even without leaks, the areas around drinkers frequently become wet and fouled and act as focal points for disease. The problem can be reduced by raising drinkers on to or over wire platforms as soon as chicks are big enough to climb on to such platforms. It is, of course, important that the chicks cannot actually get underneath. All wet patches of bedding, even when not arising from floods, should be removed and replaced periodically.

Header tanks and water pipes should themselves be cleaned occasionally - once between rearing seasons will suffice. They can be disinfected and then flushed out with clean water. It is important, too, that header tanks are kept lidded. A header tank may service a number of different brooding pens and this can make water medication of individual pens difficult. When medicating through the water, in such circumstances, it is better to cut the supply to the pen being treated and to fill the drinkers manually, preferably adding extra drinkers in the process.

As a guide, provide at least one drinker per 100 chicks started. However, if one is only rearing a batch of 100, one drinker will not be sufficient since the chicks may not find it in time. Do not provide water in rusty containers. Pheasants have a poor sense of taste for most substances, but do have an active dislike of water contaminated with rust. They will, therefore, reduce their voluntary water consumption, feed less and grow poorly.

H. CONTROL OF FEATHER PECKING.

i) General.

Chickens, turkeys and pheasants will all indulge in feather pecking under farmed conditions and it is an important part of management to ensure that this is controlled or, better,prevented entirely. Uncontrolled feather pecking will lead to vent pecking, cannibalism and such severe feather loss that, in the case of pheasants, they would die of exposure on release (*fig 9*).

The severity of feather pecking is increased with increasing light intensity and increasing stocking density. While deficiencies of certain amino acids and of salt can lead to increased aggression and feather pecking, such problems are rare although some keepers claim that tail pulling in release pens can be reduced by putting out blocks of salt in weather proof containers for birds to peck at. Mechanical damage to growing feathers can result in bleeding and the presence of blood on the feathers will certainly precipitate pecking. Once the habit starts, it will quickly spread. Therefore, if one does find a bleeding poult in the pen, it should be immediately caught. The bleeding feather or feathers should be plucked, so stopping further bleeding. Blood should then be washed off the bird before returning it to its flockmates.

The occasional group of grass reared poults, given plenty of space and the correct weather conditions, will reach release age without indulging in feather pecking. However, it should not be inferred from this that feather pecking can reliably be prevented by rearing under more natural conditions. It is, therefore, most economic and most humane to assume that feather pecking will occur unless one takes active steps to prevent it.

In the case of broiler chickens, low light intensity is used to prevent the problem. In the case of turkeys reared entirely in windowless housing, the same approach applies. However, if they are to experience natural light, beak trimming becomes essential. This is done in such a way that it is permanent. Both of these approaches can be used in pheasants but there are reservations. Poults must be used to natural light by the time of their release and must have undamaged beaks so that they peck food from the ground. It is for this reason than bits have been developed.

ii) Light intensity.

The use of low light intensity until chicks can be bitted at or soon after three weeks of age is a very effective way of preventing feather pecking. However, as has been stated previously, a very low light intensity during the first few days can increase mortality through 'starve outs' and a non-fluctuating low light intensity can lead to blindness in a few birds. As a guide, use a light intensity level during the first four days that enables one comfortably to be able to read a newspaper, but avoid bright light. Thereafter, until bitting, lower the light intensity, but always ensure that there is still more light by day than by night.

When using brooder huts and shelter pens, light control is less easy. Prevent shafts of sunlight from shining into the huts through windows or ventilation holes without interfering with air supplies. Early access to the shelter pen in bright conditions may lead to losses through wing tip pulling. It may be possible to shade the polythene of the shelter pen with hessian in such circumstances. In any event, use opaque and not clear polythene.

PHEASANT HEALTH AND WELFARE

Fig. 10
Beak tipping with an electric debeaker. The beak will regrow within 10 to 14 days if just the tip is removed. During this period feeding efficiency will be reduced and poults will have difficulty in pecking feed directly from the ground.
(Photo courtesy of the late T. Blank)

Fig. 11
Full beak trimming. Half of the top beak has been removed. This is a semi permanent operation and should never be undertaken on birds which are to be released.
(Photo courtesy of the Game Conservancy)

iii) Beak trimming.

In pheasants, it is only possible to remove a small proportion of the upper beak at one time since regrowth is essential. One should not remove more than 20 per cent of the distance to the nostril from the tip (*fig 10*). Under such circumstances, the beak will regrow in ten days to a fortnight. This is too short a period to be of routine use in control of feather pecking, but may be valuable in certain circumstances:

a. If one expects feather pecking due to bright light conditions before three weeks of age and if one does not want to apply small bits, one can trim the beaks at ten days of age and then apply medium sized bits at three weeks. This is obviously an extra chore and an extra stress to the chicks. It may, nevertheless, be necessary when using huts, night shelters and grass runs with no possibility of controlling light intensity.

b. If one has failed to bit a pen of grass reared pheasants at three weeks because no pecking has occurred, there may be an outbreak of it at, say, five weeks. In such a case, beak trimming would be as effective and more labour saving than bitting.

c. If bad weather or a delayed harvest require that poults be kept back in their rearing pens for a further week or so before release, their upper beaks will curl around the bits thus enabling feather pecking to start. One has two options - the fitting of larger bits or slight beak trimming while leaving the original bits in place. If the delay in release time is likely to be short, the latter option is the most sensible.

d. Some keepers or poult purchasers like to have poults routinely beak trimmed at the time of release to prevent problems of tail pulling. Although some tail pulling occurs in many release pens, this is generally of no consequence and tails will be fully grown by the start of the shooting season. If people experience severe problems with consequent vent pecking and tail-less pheasants in November, they should be more concerned to correct the inadequacies of their release pens or their stocking policy rather than to beak trim the poults at release. Release is a time of severe stress when feed intake often dips anyway without adding the additional stress of beak trimming. It should also be appreciated that beak trimmed pheasants cannot be expected to peck feed from the ground with any degree of efficiency and should, therefore, be trough or hopper fed until the beaks have regrown.

e. If one wishes to retain a number of poults for future breeding stock, they are easier to manage if they are given a permanent beak trimming. This requires that the upper beak is removed half way back to the nostrils. Clearly, such birds should never be released and never fed off the ground (*fig 11*).

All debeaking should be undertaken with a proper electrical debeaking device.

iv) Bits.

Although metal alloy bits are still made, the great majority of bits used are now plastic. There are two types of plastic - one soft and pliable, the other more rigid. It is now also possible to buy a tool which holds a magazine of bits and which, according to the claims of the manufacturer and many users, speeds up the bitting process.

The smallest sized bits can be applied at 10 days of age. Given adequate control of light intensity, it is not necessary to bit this early. However, as has been stated previously, not everyone is in a position to control light intensity. Although some people claim that a single bitting at 10 days is all that is needed, most will find that the upper beaks close over the bits in a few weeks, rendering them ineffective well before release is due. One is then forced to catch all the birds a second time to fit larger bits or to beak trim. This is extremely time consuming and tedious as well as stressful to the birds. It is generally better, therefore, unless one has a severe pecking problem, not to bit until three weeks or soon thereafter. At this age, the second sized bit can be inserted and its beneficial effects will last certainly for three weeks and maybe for up to four (*fig 12*). By this time, the great majority of poults will be due for release.

The two larger sizes of bit are seldom used unless, for some reason, one wishes to hold birds for longer than seven weeks. However, the prolonged wearing of bits can lead to permanent beak deformities and there is a strong case to be made for replacing large bits with small sized clip-on 'spex' which are less damaging to the beak and just as effective in preventing feather pecking.

When applying bits at around the age of three weeks, one has the choice of metal, soft or hard plastic. Metal bits are slower to insert and require to be cut off. They have, therefore, nothing to recommend them. Soft plastic bits are, for the uninitiated, easier to apply, easier to remove and more liable to fall off than hard plastic ones. However, with practice, the stiffer bits can be applied and removed just as quickly. Bits represent a not inconsiderable cost. Hard plastic bits can be used many times. The softer ones quickly lose resilience and, the more frequently they are used, the greater the proportion that fall off.

The second sized bits can be applied as early as eighteen days but experience suggests that it is far less stressful, unless there is an already well established pecking problem, to bit at some time between 22 and 25 days. The process of catching and bitting does come as a shock, albeit temporary, to the chicks. It also takes them time to adapt to the bits, and feed intake and growth rate will be temporarily checked. If there is subclinical disease in the flock, bitting can precipitate a disease problem, particularly coccidiosis due to reduced intake of anti-coccidial drug. Chicks will also feel the cold more due to their reduced appetites and are apt to huddle together with increased risk of suffocation in 'pile-ups'. One should never bit undersized chicks in the

REARING

Fig. 12
A chick fitted with a bit. While a bit will prevent feather pecking until the beak has grown around it after approximately three weeks, feeding is made more difficult. It is thus necessary to ensure that there is a good depth of feed to peck at. One should not change from crumb to pellet feeding at the same time as bitting the chicks.
(Photo courtesy of the Game Conservancy)

Fig. 13
Three day old chicks in a gas brooder unit. The feed is presented in base pans of tube feeders. The tubes will be added after a further couple of days and the feeders will then be suspended. The surround (ring) will be removed at the same time.

flock. It is best to catch them up later when they have increased in size. If the attendant always carries a few bits in his pocket, he can catch such birds as he comes across them and bit them at an appropriate time.

It must be appreciated that the wearing of bits makes feeding more difficult. The process of bitting should not coincide with a change from crumbs to pellets. Recently bitted birds can cope adequately with small mini-pellets provided that they are used to them well before the time of bitting. However, they cannot deal with the larger mini-pellets that are produced by some compounders. In the latter circumstances, it is best to feed starter crumbs until two or three days after bitting and then to change gradually over a further two to three days to grower mini-pellets. Regardless of whether one is feeding crumbs or pellets to bitted birds, the feed must be presented in feeders that provide a decent depth of feed. A thin layer cannot readily be picked up.

Some rearers regularly experience mortality of one to two per cent in the 48 hours after bitting. This is unacceptable and indicates a management fault (or faults) that should be corrected.

I. DAILY MANAGEMENT.

Much of this subject has been discussed in earlier sections. However, there are two areas that have not been covered.

i) Introducing chicks to the brooder and the early days.

Contrary to what has often been written in the past and what is quite widely believed, chicks should be placed in their brooders as soon as possible after they have been taken from the hatchery. Equally, they should not be kept in the hatchery for very long after they have fluffed up. While it is true that chicks are hatched with yolk reserves of nutrients that will keep them alive for three days without feeding, it is equally true that the longer they are kept in chick boxes the poorer will be their subsequent performance. There is a direct relationship between the length of time between hatching and placement under brooders and the number of chicks that 'starve out' between the fourth and sixth days. For every chick that actually dies during the 'starve out' period, there are almost certainly several more that do not die, but get off to a bad start.

The brooders should be heated up to optimum temperature several hours before chick placement. Care, however, should be taken not to overheat. This can easily occur in brooder huts with gas brooders, particularly when the ventilation is restricted. The chicks should initially be confined within rings in the brooder area (*fig 13*). The confining rings should be circular or oval and made of hardboard, fibre glass, corrugated paper or even, in the guaranteed absence of draughts, of fine mesh wire netting. They should be placed in such a way that the heat source is not aimed at the centre of the ring. If the ring is slightly offset, there will be a temperature gradient across it, giving the chicks more choice of finding a correct comfort zone. Obviously, after an hour or so, one should check the behaviour of the chicks within the ring. If they are heaped under the point source of heat, they are too cold. If they are pressed round the edge of the ring, the area beneath the heat source will be much too hot. If the whole environment is too hot, the chicks will look torpid and sit around singly or in groups with their wings drooped. It is a common fault, particularly with gas heaters, to overheat the entire area of the ring. While it is acceptable to have a temperature of slightly over 100°F immediately beneath the heater, it is important that the temperature at the periphery of the ring be no more than 90°F. If one were to reduce both these figures by 5°F, no harm would be done. However, to increase them by the same amount would be dangerous.

As the chicks age, the area within the ring should be expanded until, by the fifth or sixth day, the ring should be removed altogether. If a low ring is used, one should be careful that chicks don't fly out prematurely and get chilled. Corrugated paper rings can easily be knocked and displaced allowing chicks to escape beneath.

Shortly after removing the ring, return to observe the behaviour of the chicks. It is at this stage that cold draughts may lead to huddling and smothering. Even in the absence of draughts, the removal of the ring often disorientates the chicks which may then seek out and heap in pen corners for security. It is for this reason that corners should be rounded off. In the case of huddling, correct the cause of the problem if it is a draught and stir the chicks up. If they still persist in heaping where not required, one may have to move heaters, feeders and drinkers around until the problem is solved.

ii) Hardening Off.

Regardless of the rearing system used, considerable management skills are required when chicks are first given access either to shelter pens, grass runs or outside covered runs. Environmental temperatures and wind directions can change quickly and the attendant must be prepared to shepherd birds between inside and outside areas according to conditions. For the first few days after chicks are let outside by day, they should be shut up at night or sooner if they look cold or miserable. Brooder house temperatures should be gradually reduced until no artificial heat is provided at all. This is usually at about four and a half weeks, but will very much depend upon housing system and outside temperatures. Well before this, heaters can often be turned off completely by day and on again at night. Poults should never be released until they have been totally weaned from artificial heat for at least ten days.

After several days of shepherding chicks back to bed at night, it is normally possible to leave popholes open so that the chicks can make their own arrangements. At this stage, if there is an internal light left on until after dusk, chicks will be more likely to return to the security of the brooder pen than to jug outside. The light should, however, not be left on all night. At the late rearing stage, it won't matter if the poults do jug outside except, possibly, in cases of torrential rain or in abnormally cold conditions. Under such circumstances, it will pay to visit the outside pens to see if birds are heaping and smothering or chilling off in long wet grass. If there is any sign of this occurring, it will be worth stirring up the poults and pushing as many as possible under cover even if this does create a degree of panic in the darkness.

When birds are reared entirely under cover they will never have been rained upon. Even if they do have access to grass runs, the same may apply in dry weather. It has been demonstrated conclusively that, if the feathers are well wetted on three or four occasions during the later stages of rearing, poults are much better able to withstand thunderstorms or heavy rain at or soon after release. It will, therefore, be of advantage to spray birds in covered pens with water from a knapsack sprayer or other device on a number of occasions in their last ten days of rearing. If one starts this procedure too soon, one will create wet litter conditions conducive to the rapid development of coccidial oocysts. The same procedure can be adopted in grass pens if it has not rained. If it does rain, it may be worth shutting poults out of covered areas for a period to ensure that they all get wet.

J. CATCHING FOR RELEASE.

i) General.

Several procedures may require to be undertaken when catching and crating poults to take them to their release pens. These will obviously include de-bitting when bits are worn and may include wing clipping, beak tipping and poult identification. It is, therefore, helpful to have several members of a catching team so that the whole process can take place quickly. It is desirable that a period of good weather is chosen for release, but this is not always possible and, so long as the poults are well hardened off and the release pen contains good cover, only very bad conditions are likely to be of consequence. In hot weather try to make a very early start, but never crate the birds the night before delivery. Regardless of weather, it is desirable to stock release pens earlier rather than later in the day so that poults have plenty of daylight during which they can learn something of their new surroundings before nightfall.

Crates should never be overfilled. If birds are older than normal when they are crated, make allowances for their increased size and reduce the numbers put into each crate. In crates with solid floors, add shavings to absorb faeces and to provide more grip. Those with slatted floors should not be stacked directly on top of each other without a paper sack or something of the sort between, otherwise poults in the lower crates will receive a rain of faeces from above. If possible, do not crate wet poults because they will be subject to feather-loss along their backs if they climb over each other. This is liable to happen to some extent anyway, particularly if the crates are tipped or if they are too high in relation to bird size. When stacking crates into a trailer, ensure that plenty of air can circulate between stacks to avoid overheating.

ii) Culling.

Not all birds in a pen will be suitable for release at one time. A few may be undersized or feather-pecked and should be retained in a separate pen till better grown or feathered. A small number of others are usually better culled than released. These may be made up of blind birds, those with crooked necks or legs, those with humped backs, and those pecked on the wing joints with subsequent joint swelling due to infection.

While it is true that birds with deformed legs and necks may well survive to the shooting season and fly respectably over the guns, such poults should never be sold. Even if they are to be released in an estate's own coverts so that the question of sale does not arise, they are a poor advertisement for the shoot and are best culled as poults.

iii) Clipping or Pulling Wings.

If it desired that poults should not immediately fly out of open-topped release pens, then some of the primary flight feathers on one wing can be cut or pulled out. One should only ever cut those primary feathers which are shortly due to drop out anyway. If one cuts new feathers, they will not be replaced until well after the shooting season has ended. If the wing is extended, those that can be cut will be found furthest from the body and the distinction between old feathers about to be replaced and new ones, closer in to the body, will be obvious. There are ten primary flight feathers and the numbers that can be cut will depend upon the age of the bird (*fig 14*). At six to six and a half weeks, six or seven may be cut. By eight weeks, only three or four have not been replaced and even these will be about to drop out anyway.

It follows that, if birds are released much after seven weeks, wing clipping is unlikely to keep them in the release pens for very long. One may, therefore, have to consider wing pulling. This involves the actual plucking of seven or eight primary feathers, most of which will be new and some of the quills of which will contain blood. This must be done with care and is a much slower process than wing cutting. Even when carefully undertaken, it is likely that some damage will be done to feather follicles and, although replacement feathers will appear, some may be twisted and defective. This may well impair flying ability. If wing pulling is done roughly, the consequences can be more serious with a proportion of poults suffering dislocations of their terminal wing joints.

Fig. 14
Wing clipping prior to release. The outermost primaries that have not yet moulted should be cut. The innermost primaries are new feathers which have just grown out and, if cut, will not regrow that season.
(Photo courtesy of the Game Conservancy)

iv) Poult identification.

There are three ways of marking birds at release so that they may be subsequently identified:

a. Wing tags. Numbered tags, available in several colours, can be purchased. The tag is inserted into the leading edge of the wing in the wing web at the elbow joint with a pair of special pliers. When applying tags, allow room for growth and do not insert them too far into the web. Properly applied, only a small percentage of tags should be lost.

b. Leg rings. Leg rings, consisting of coiled flat plastic, are available in a variety of colours. These tend to be more expensive than wing tags, take longer to apply and are more likely to be lost.

c. Toe web slitting. The loose web of skin between the toes can be cut to about one third to half its depth with a sharp pair of scissors. This will rapidly heal with minimal bleeding, but each edge of the cut will heal independently so that a permanent slit or notch is left which can always be subsequently identified. Since there are four webs, inner and outer on each foot, several different identities can be established, particularly if more than one web on any one bird is slit.

This is a cheap and quick method of identifying birds and is more foolproof than other methods. However, confusion can arise if the practice is adopted by several neighbouring estates without prior consultation.

K. PROBLEMS.

i) The first two weeks.

a. Minor mortality during first three days.

It is usual to expect to lose a few chicks during the day or two after placement. Such chicks will probably have suffered yolk sac infections by one of a number of bacterial agents. The infections may have been introduced to the egg soon after laying, or may have got in through the navels of chicks on the hatching trays. When buying chicks, it is normal to expect an extra two per cent free of charge. This should, in normal circumstances, more than cover the losses. Losses will be somewhat higher when second quality chicks and those helped from the egg are brooded. In addition to greater losses through yolk sac infections in such chicks, there will also probably be a few that fail to find their feet and quickly die of dehydration if not culled. A few chicks may fall into drinkers and drown at this stage.

b. Moderate or heavy mortality during first three days.

Losses of this type are usually due to serious management faults. The commonest cause is probably overheating, particularly likely in gas-heated brooder huts. If one skins the corpses of chicks killed in this way, the flesh will appear very pale. Losses through chilling are much less common, but may occur in cases of brooder failure as when electric supply is interrupted or gas runs out.

The other likely cause of major mortality at this stage is dehydration through failure to find water. Even when water is available, chicks may fail to get to it if it is sited in a cold or dark area relative to the heat source. This is most likely to occur when brooding in cold conditions with electric hens as the only source of heat or when there is inadequate light in the first couple of days.

c. Mortality during the fourth to sixth days.

A small, or occasionally large, spate of mortality during this period followed by its abrupt cessation is indicative of a 'starve out' problem. Birds dying at this stage have generally never fed and will have empty guts and enlarged gall bladders. The extent of the 'starve out' problem is related to management. If chicks are taken from the hatchers as soon as possible after they have fluffed up and immediately placed in good brooding conditions with easy access to well-illuminated feed and water, there will be little or no mortality. Many keepers expect to lose in the region of four per cent of chicks at this stage and seem quite unaware that the solution to the problem lies in their own hands. This is usually because they mistakenly believe that they are actually benefiting chicks by holding them in chick boxes for prolonged periods before placing them under brooders.

d. Moderate or heavy mortality continuing beyond day six.

The causes of such problems are more likely to be disease associated than due directly to poor management. In such circumstances, veterinary help should be sought. Some keepers have access to illicit supplies of anti-bacterial agents and use them at the first sign of trouble. While they may occasionally nip problems in the bud in this way, more often they will do more harm than good. As an example, furaltadone and furasolidone are useful drugs for pheasants, but they are also potentially extremely toxic. Many pheasants die every year through 'furasol' toxicity.

The main bacterial cause of mortality at this stage is *Escherichia coli*. *E. coli* are always present and there are many different types, some potentially more harmful than others. However, they generally do no damage unless present in large numbers or unless they gain entry to chicks at a very young age, particularly if the chicks have been stressed, chilled, overheated or kept too long in chick boxes. A certain number of eggs will always produce chicks that are infected with *E. coli* at hatching. These can infect other chicks on the hatching trays or in the chick boxes. If they are infected via the navel, they will probably die early with yolk sac infection. If they inhale the bacteria, they may subsequently develop septicaemia and tend to die between the third and tenth days. Unfortunately, many strains of *E. coli* are resistant to many of the more commonly used anti-bacterial agents and sensitivity tests will be necessary before a suitable drug can be reliably found for treatment. It should be appreciated, however, that, while it is *E. coli* that is actually causing the chicks to die, it is usually poor management that produces the situations in which such mortality is likely to occur.

Various species of salmonella can cause more or less the same types of disease pattern and signs as *E. coli* at this age. The vast majority of salmonella species are harmless to birds that are infected after five days of age. However, if significant numbers of such bacteria gain access to the guts of chicks in their first two days of life, they may multiply, escape into the blood stream and cause septicaemia. Salmonellosis as a disease is thus associated with very early infection which usually means that it arrives with the chicks from the hatchery. The bacteria gain access to the hatchery through faecally contaminated eggs and, if hatchery hygiene is poor, may persist for weeks or months in the hatchery environment. However, the presence of salmonella in chicks leaving a hatchery does not imply that such chicks will necessarily suffer. Undoubtedly, infection will spread rapidly from chick to chick. However, if the numbers of bacteria are low and the chicks are not stressed, there may be no disease. Very often, a hatchery may send out several batches of chicks, all infected with salmonella, and only one batch will suffer significant mortality. Investigation will usually reveal that the group which suffered was exposed to a long journey or was chilled or otherwise stressed. Some strains of salmonella are more likely to kill chicks than others and, on rare occasions, mortality may reach 70 per cent. In such circumstances, it is probably wise to cull the survivors as they will all be carriers. If possible, never use pheasants which have experienced salmonellosis as breeding stock.

It will hardly have escaped anyone's attention that the presence of salmonella in poultry has caused much political excitement. Salmonella is a major cause of food poisoning and it follows that anyone dealing with salmonellosis in pheasants should take hygienic precautions. If a veterinary surgeon diagnoses salmonellosis or even grows salmonella from a chick that may have died of something quite different, he has, by law, to report the fact to the Divisional Veterinary Officer of the Ministry of Agriculture.

In recent years, rotavirus infections have been increasingly diagnosed as causes of quite severe mortality in pheasant chicks.
Affected chicks become stunted and are obviously smaller than their fellows after approximately five days. Over the next week, there may be considerable mortality. Affected birds tend to have frothy bloated caecal contents which must cause them considerable discomfort so that they feed little and have a very hunched appearance. There is no specific treatment for the problem although some veterinarians may prescribe anti-bacterial agents to control secondary infections, or vitamin and electrolyte mixtures to help overcome the consequences of scouring. Disease caused by rotavirus tends to be seen later rather than earlier in the season. This implies that it can probably be prevented or greatly reduced by improved hygiene so that chicks are not exposed on placement to a highly contaminated environment. While the virus is not believed to be transmitted through eggs, it is possible that it can enter hatcheries on the surface of contaminated eggs or the clothing or hands of contaminated personnel.

Very occasionally, pheasant chicks suffer from so-called brooder pneumonia due to the fungal infection, aspergillus. The problem can be egg-transmitted or can arise if chicks are exposed to mould spores in the environment. When grass in runs is cut and left lying on the ground, it may go mouldy and cause the disease. Even if it doesn't, it is likely to lead to impacted gizzards, so if it is considered necessary to cut grass, ensure that it is picked up and removed. Chicks affected with aspergillosis gasp for air, sometimes arching their necks, and a proportion die. Others may develop a dry cough. There is no economic treatment, but the problem can be prevented by good management.

ii) The third week to release.

a. Coccidiosis.
When rearing birds in large groups, one must expect that they will be challenged with coccidiosis, however good the management. Good hygiene will reduce, but not eliminate, the level of challenge. It is, therefore, extremely foolish to use a ration that does not contain an effective anti-coccidial drug.

There are currently only two anti-coccidials that are licensed for use in pheasants and there is quite widespread resistance to one of these. If resistance starts to develop to the other and if no new drugs become available, it will become necessary to switch between the two from season to season in the reasonable anticipation that the parasites will lose resistance over time if they are not continuously exposed to a drug.

There are three main species of coccidial parasite that affect pheasants in the U.K. These are host specific and thus quite separate from the coccidial parasites of other birds. In other words, pheasants can't be infected by contact with partridges, chickens or turkeys, etc. One of the species, *Eimeria colchici*, attacks the blind guts or caeca. The others are found in the small intestine. Caecal coccidiosis tends to be more severe than intestinal and may occasionally be seen as early as ten days of age. However, coccidiosis is normally encountered from two weeks of age. There are no specific signs other than birds looking miserable and going off their feed. Diarrhoea is not necessarily a feature and, in some types of disease, there may even be constipation. The level of mortality, which can be high, will depend upon the degree of challenge, the extent to which active immunity has built up and the efficacy of the anti-coccidial drug. As has been stated previously, the level of drug included in the diet is such that it should allow some coccidial development and thus allow immunity to develop. In the case of very heavy challenge, therefore, the level may prove inadequate to prevent disease. Conversely, if there is too much drug, no immunity will develop and, when birds are released and cease to receive the anti-coccidial, they will be totally susceptible to the infection. Any other disease or management practice, such as bitting, which interferes with feed consumption, will reduce anti-coccidial intake and may thus lead to secondary coccidiosis.

Coccidiosis is passed from bird to bird via the oocyst which can be regarded as the coccidian equivalent of an egg. Vast numbers of oocysts can be passed in the faeces. They are extremely tough and are not killed by most disinfectants. They are destroyed by ammonia, but this is very unpleasant and dangerous to use. It is not usually practical to attempt to kill oocysts by disinfection although some manufacturers claim to produce effective compounds which, on combination, will release ammonia. While routine disinfection is not generally worthwhile, it must be emphasised that the reuse of litter and the reuse of grass runs within a given season are extremely risky practices. The oocyst is not immediately infective when it lands on the ground. First, it has to sporulate which takes a variable time depending upon conditions. A hot, damp environment will enable sporulation to develop very quickly. Wet litter should definitely be avoided and fouled litter around drinkers should thus be replaced regularly.

All drugs used for treatment, as opposed to prevention, of coccidiosis are classed as Prescription Only Medicines (POM) and can, therefore, only be obtained on veterinary prescription.

b. Mortality soon after bitting.

Some pheasant rearers expect to lose two or three per cent of chicks within a few days of bitting them. This should not occur and indicates some sort of management problem or management/disease interaction. The normal age to bit pheasants is approximately three weeks. Premature bitting with size 2 bits will increase the likelihood of mortality. It is generally better to bit a few days after, rather than a few days before, three weeks of age. If birds are undersized or uneven in size due, for example, to slight but undiagnosed coccidiosis, bitting should be delayed further unless feather pecking is already a serious problem. The importance of not changing from crumbs to pellets just before, or at, bitting and the need for a good depth of feed in feeders after bitting has previously been emphasised.

If birds have slight coccidiosis before bitting, the reduced feed intake immediately following bitting will lead to less ingestion of anti-coccidial drug and thus severe coccidiosis can be precipitated. One is often trying to harden chicks off at around this time. Bitting stress makes chicks feel the cold and they often huddle in heaps with resultant suffocation of some birds. This phenomenon, too, is made worse by subclinical coccidiosis.

c. Respiratory Disease and Septicaemia.

Many of the rearing systems used for pheasants work well under normal conditions for a single batch at a time, but fail when several age groups are together or when the weather conditions are adverse. Under the latter circumstances, even when management is otherwise efficient, one is likely to experience respiratory disease and septicaemia in one's birds due to *E. coli*. There are always *E. coli* in the environment and in the intestines of birds. Huge numbers of certain types can build up in the environment and be inhaled by chicks. This is most likely to occur in overcrowded and poorly ventilated conditions. Since numbers build up anyway as birds get older, the exposure of young chicks to the air breathed by older ones or the dust created and left behind by older ones makes the problem more likely on multi-age sites later in the season. The importance of ventilation increases as the weight of birds in a given airspace is increased.

Many systems work on the assumption that chicks will have rapid access to outside runs. The inside brooding areas often have no mechanical ventilation. In very cold or windy conditions, chicks may refuse to use outside runs or may chill off and die if they do. Under such circumstances, they will crowd in dusty, poorly ventilated conditions in the brooder area and inhale bacteria. The dust and the build up of ammonia will cause damage to the lining of the trachea (windpipe). This will aid bacterial attack and lead to coughing and sick birds. Variable numbers will then die as bacteria get into the bloodstream and cause septicaemia. Under these circumstances, veterinary help should be sought and some sort of antibacterial agent will be prescribed. Unfortunately, many strains of *E. coli* are now resistant to many of the commonest antibiotics so some time may be wasted while sensitivity tests are conducted in a search for a suitable drug. Obviously, total reliance should not be placed on treatment. One must also look to an immediate improvement in air flows through affected pens without causing undue draughts and chilling. Consideration should be given to how the housing system and management can be altered in such a way as to prevent the occurrence of the problem in a subsequent season.

There is another respiratory problem that appears to be on the increase in pheasants. This is caused by the agent, *Mycoplasma gallisepticum*, which produces a chronic cough and sinusitis. Occasionally, other mycoplasma species are involved. The sinuses below the eyes swell, causing the eyelids to close and the eyes to water. The chicks may be seen rubbing their heads and eyes. The infection spreads directly from bird to bird and can also be transmitted through eggs so that new generations become affected. While it is possible to treat flocks quite successfully to prevent the spread of the clinical disease to flockmates and even to cure the coughs of affected birds, it is difficult to cure the sinus damage and almost impossible to eliminate the infection. Even treated birds will remain carriers for very long periods. Such carriers may not spread much infection when they have left the release pen, but, if caught up the following spring for breeding, they will still be capable of infecting other laying stock as well as transmitting disease through their eggs. Therefore, once the problem appears on an estate, it is likely to persist from season to season. This need not necessarily result in significant mortality so long as strategic medication is routinely given. However, it is clearly desirable to take drastic action when possible to prevent the establishment of the disease in the first place. Therefore, if the signs first appear in a small batch of chicks, it may be desirable to kill them all, regardless of whether they are all obviously diseased or not. In many circumstances, this may not be an economic proposition. In any event, expert veterinary advice should be sought as soon as possible. If one must release affected groups, they should be tagged and subsequently discarded from the next season's breeding programme. If the disease has become persistent on an estate, it is possible but very difficult to eradicate it.

d. Hexamita and Trichomonas.

The two very similar diseases of hexamitiasis and trichomoniasis are caused by protozoan parasites which, when present in large numbers, cause intense scouring with very rapid weight loss, dehydration and death. They are normally encountered in release pens, especially in wet weather and will, therefore, be discussed more fully in the following chapter. However, the diseases can occur pre-release in grass reared poults, particularly when the grass has been contaminated by a previous batch of poults. Under such circumstances, mortalities of over 70 per cent can occur if there is no preventive drug in the ration. More often, there are only smaller numbers of parasites present and the poults will appear healthy but will be carriers. They may then break down with disease later in the release pen if adverse circumstances occur or they may seed the ground with infection which can be picked up by previously healthy, but susceptible, birds.

REARING

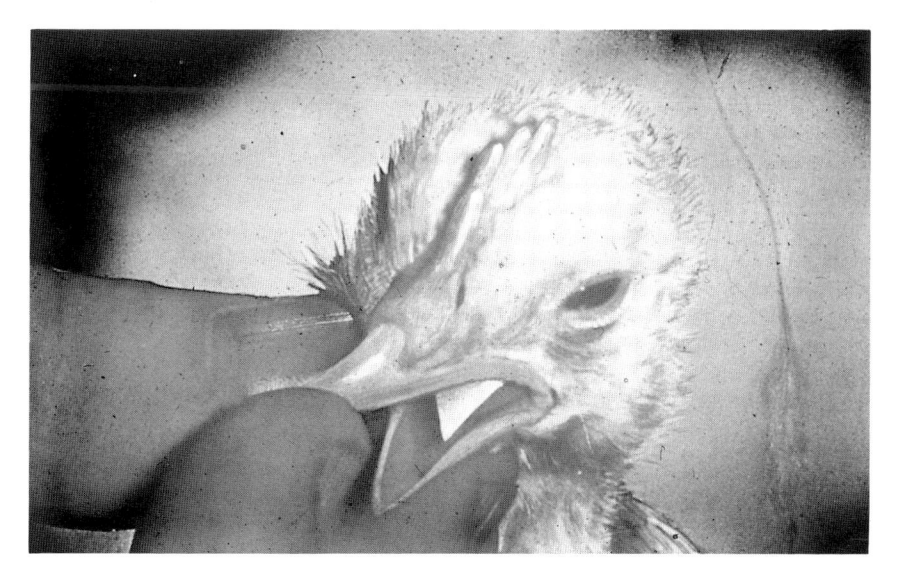

Fig. 15
A chicken with rickets. This condition can occur in pheasants and is often first noticed at bitting when the rubbery beaks make the application of bits difficult. If rickets is suspected, change the feed, contact the feed supplier and contact your veterinary surgeon.

e. Worms.

Intestinal worms and gapeworms can cause problems during rearing, but, as above, are much more likely to do so in release pens and will thus be discussed in the following chapter. Again, it is only in grass reared birds that the problem is likely to arise during rearing. Worm eggs can survive in the environment for quite long periods and it is thus important to rear birds on grass that has not been contaminated with faeces of adult birds during the previous shooting season.

It is commonly believed that grass reared poults will be superior after release to those reared in other systems. This is often not the case and it should be appreciated that such poults are more likely to be sufferers or carriers of protozoal and worm infections than are those reared under cover. If the same grass is used for two consecutive batches of chicks within a single season, there is a greatly increased hazard of resulting problems.

f. Nutritional problems.

Problems arising from feed compounder error are uncommon nowadays. The most frequently encountered of such problems is rickets, a condition in which the bones and beak do not harden properly (*fig 15*). Birds affected with rickets will sit around and be reluctant to stand up. When they do, they tend to tremble and shake. Rickets is often detected first at bitting. The beaks are soft and rubbery thus making it very difficult to apply the bits. The problem can be rapidly corrected, if diagnosed early, merely by changing the diet. Another, albeit very rare, condition caused by dietary deficiency is chondrodystrophy in which significant numbers of poults develop crooked legs which angle out or sometimes in at the hocks. If only a few individuals are affected, the feed cannot be blamed.

Very occasionally, a feed may contain a wrong drug inclusion which can prove damaging to pheasants. By the same token, the declared drug might have been accidentally omitted or put in at too low a level. Mislabelling may result in a feed of too high or low a protein level being used with adverse consequences upon growth. Rarely, feed may be contaminated with fungal toxins which can lead to stunting and increased mortality. Stale or badly stored feed can also result in problems associated with rancidity and vitamin loss.

Whenever a feed problem is suspected, the feed compounder should be informed at once. Most compounders are ethical and, if a genuine problem is proved, will compensate without question. However, to be on the safe side, one should also take one's own feed sample, preferably in the presence of the compounder's representative, for storage. This can be sent for independent analysis later and only if necessary. It is expensive to have feed analysed and one must have a very detailed idea of what to ask the analyst to look for. If a dispute with the compounder appears likely, it will probably be worth taking veterinary advice. There is a defined method of sampling feed which is legally acceptable. Failure to adhere to this method can weaken one's case should any dispute go to litigation.

iii) Summary of Rearing Problems.

The previous sections are by no means comprehensive, in that only the commoner problems are mentioned. However, it should be clear from what has previously been written that diseases are not generally 'Acts of God', but are much more likely to be direct consequences of rearing system or management failure.

If an employer wishes his keeper to rear pheasants, it is his duty to ensure that the keeper is provided with the means to do so. If the employer doesn't understand the requirements, as is often the case, he must either put his faith entirely in the keeper and provide what is asked for or he must seek independent advice. What he is not entitled to do is to demand a quart from a pint pot. If the rearing system is not well designed and equipped, management is made infinitely more difficult and, sooner or later, there will be diseases or other disasters.

Even with a good system, an ignorant keeper still has many opportunities to produce bad results. He may, for example, hold chicks in boxes overnight before putting them under brooders, greatly increasing the chances of 'starve out' problems and even of salmonellosis if the chicks should happen to the infected with salmonella. His bitting practices may fail to prevent feather pecking or may precipitate coccidiosis. Lack of attention to detail during hardening off may lead to large losses in 'pile ups'.

Given a good system and a good keeper, it is still necessary to have an efficient anti-coccidial drug in the feed and to start with healthy day old chicks. Several infectious agents can arrive on the rearing site with the chicks. Good management will reduce the level of disease associated with such infectious agents, but cannot be relied upon to eliminate their impact altogether.

L. TARGET PERFORMANCES.

i) Mortality.

If one buys in day old chicks, one would expect them to have been graded at the hatchery and thus they should all be of first quality. One would further expect to receive two per cent extras. Under such circumstances, one should aim to take no less than 95 per cent of the total started to the release pen - or 97 per cent of the number paid for. The 5 per cent loss should be the target figure for the whole season and allows for an occasional problem batch. If rearers routinely expect to lose somewhere between 10 and 15 per cent of chicks during rearing, there is something drastically wrong with their systems or their management and they should take advice on possible corrective measures.

When one is rearing second quality chicks and those helped out of the shells, one should expect increased mortality and an increased cull rate. One's target should be to lose no more than ten per cent of such chicks during rearing, assuming that reasonably rigorous culling has been undertaken in the hatchery before placement of the chicks under the brooders. Thus, if one rears a batch of pheasants composed of 80 first quality chicks and 20 second quality, one's target or expectation should be to end up with no less than 94 good quality poults.

ii) Weight Gain.

A graph is shown which demonstrates the growth rates of cocks and hens of a light to medium strain of game farm pheasant (*fig 16*). The birds were grown at or near maximum rates. True wild pheasants would be expected to be up to ten per cent lighter at any age, and some game farm strains up to twenty per cent heavier. It has previously been stated that heavy strains are not desirable in that they are likely to fly less well and they will eat more feed, the extra cost of which will be unlikely to be reflected in the prices paid by game dealers.

One must always expect a greater size variation within a batch of pheasants than within an equivalent batch of commercial poultry. A great degree of size disparity, however, particularly when the smaller birds have a

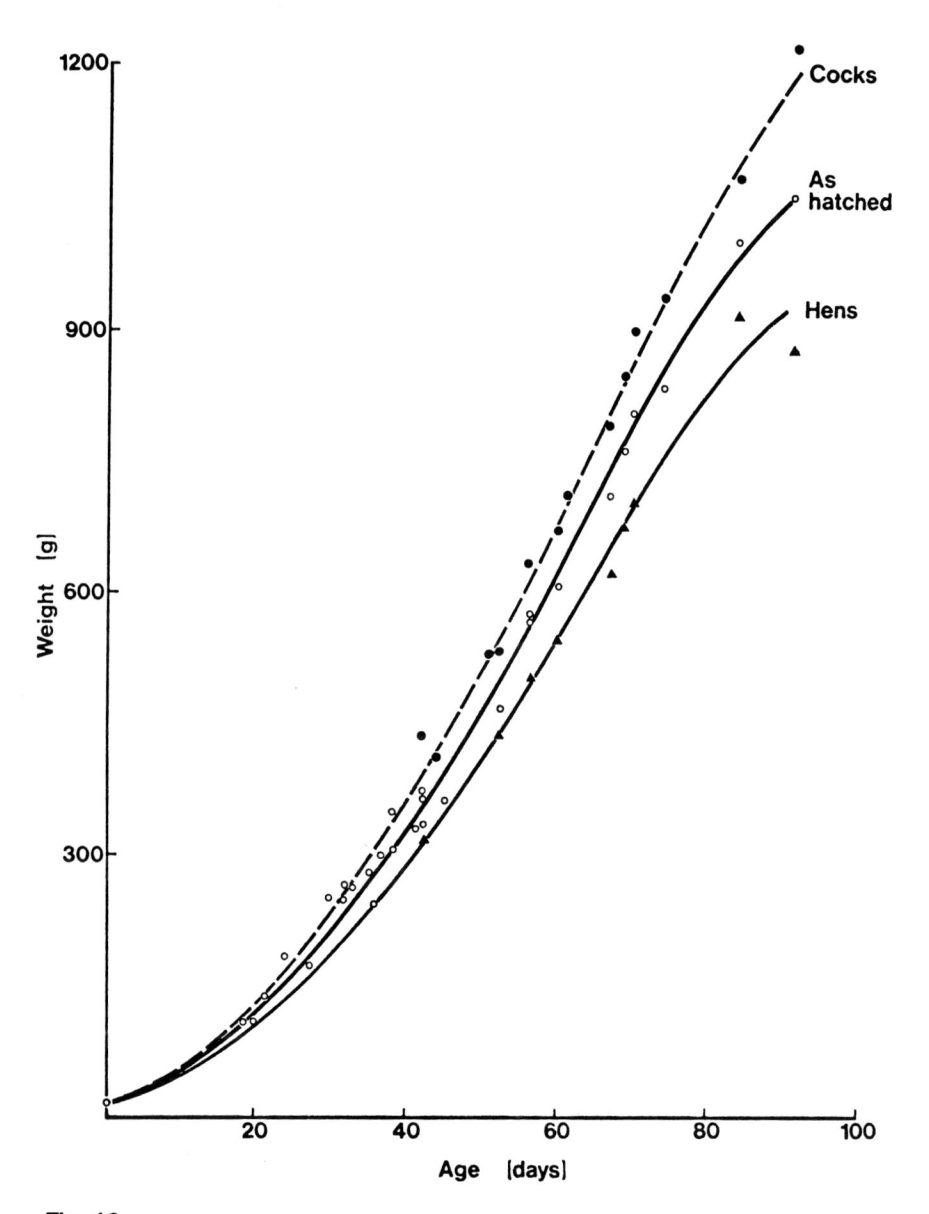

Fig. 16
Typical pheasant growth curves.

poorer body condition, suggests a disease or management problem. Poult purchasers should ensure that they handle a proportion of the pheasants as they are taken from their crates and put in the release pens. If there are significant numbers of thin birds, the purchaser should inform the supplier of his displeasure and insist that a sample of poults is despatched for post mortem examination even if the remainder of them have to be released. This can help to resolve subsequent wrangles should problems occur in the release pen. If birds are delivered which are seen to have sinusitis and if the problem is not already endemic on the estate, there is a strong case for refusing to accept any of the poults from that supplier, whether affected or not.

iii) Economics of Poult Production.

The costs of producing a poult are hugely variable. However, for what they are worth, some guideline figures are given in the Table below. These relate to delivered poults of six and a half weeks of age quoted in 1992 prices.

	Average Cost (p)	Range (p)
Chick	78	75-82
Feed	28	24-30
Labour	23	18-30
Heat, bedding and bits	14	10-20
Depreciation, interest and maintenance	30	22-40
Delivery	3	1- 5
Total	176	150-207

When costing the chick price, the potential selling price of the chick has been used, not its production cost. This is necessary when attempting to look at the viability of the rearing programme in isolation. An allowance has also been made for mortality since, obviously, more chicks will be started than poults produced.

Labour costs are extraordinarily variable and will depend upon rearing systems and numbers reared. As a rule of thumb, allow approximately 50 hours of labour per 1000 poults produced.

Depreciation costs, interest charges (real or notional) and repairs and renewals are again very variable. In general, however, they tend to be underestimated by those initially planning to rear poults. It must be appreciated that rearing accommodation has extremely limited seasonal use.

CHAPTER 4
POST RELEASE MANAGEMENT

A. THE RELEASE PEN.

i) Siting and Construction

The subject is well covered in Game Conservancy publications. Suffice it to state here that, on most shoots, there are surprisingly few good release sites. One must consider the position of the pen in relation to the proposed drives and site it in habitat that is attractive to pheasants. A surprising number of people attempt to release pheasants on parts of their shoots where there are normally no pheasants to be found. They are then disappointed that the birds evacuate the area of the pen as soon as they are able to do so.

It is also desirable that there should be vehicular access to the pen, but it should never be in a position which makes it conspicuous to the general public.

The height of the perimeter wire should be a minimum of 2 metres. It should be well dug in at the base and there should ideally be an outwardly directed floppy overhang at the top (*fig 17*). Acute angles at corners should be avoided. There should be electric fencing, preferably two strands, around the perimeter of the pen. This achieves far more than keeping mammalian predators out of the pen itself. More important, the electric wire will deter foxes from gathering outside the pen awaiting poults as they fly over the wire first thing in the morning. Anyone who has observed a dog's reaction to an electric fence will appreciate its deterrent effect.

All pens should have a sufficiency of popholes with external wings and internal funnels. These should ideally be no further than 20 metres apart and the openings should be via fox proof grids.

ii) Habitat within the pen.

Best results require that the pen is partly or, preferably, fully built within woodland. It should be reasonably draught-proof and contain tree and shrub

Fig. 17
Constructing a release pen. Note wire turned out at ground level will be pegged down and netting at top will be bent out from top straining wire to create an anti-fox fringe. Popholes will be inserted at intervals and the pen should, ideally, be surrounded with a two strand electric fence. *(Photo courtesy of the Game Conservancy)*

cover. It should also possess open areas where poults can sun themselves and dry out after rain. There should be plenty of natural low perching sites suitable for access by wing clipped poults. This will encourage the birds to become safer from ground predators more quickly and will discourage tail pulling.

Low tree branches or tall shrub cover close to the perimeter wire will lead to premature escape of wing clipped poults. In small pens, it is difficult to provide enough clear space without demolishing too much of the internal cover. For small releases, therefore, one might wish to consider a roofed pen or the release of birds with full wings. One must accept that the latter will escape prematurely, but as they are likely to do this even with clipped wings, they will be less compromised when they encounter predators.

iii) Size of pen in relation to numbers stocked.

An often-used rule of thumb is to allow one yard of perimeter wire per poult released. This is a totally unsatisfactory guide as the area per bird provided will vary enormously according to the shape of the pen and the size of the release. It is, therefore, better to think directly in terms of area per bird rather than in length of perimeter wire.

Game Conservancy research has shown that, if one wishes one's poults to have no significant adverse environmental impact to the flora within the pen itself, one should not exceed a stocking density of 240 birds/acre of pen (600/hectare). In an ideal world and with money no object, this may be an optimum stocking rate and would certainly reduce the risk of mass predator kills and tail pulling. However, although from most points of view, a generous space allowance is desirable, it does have some disadvantages quite apart from cost. In very big pens, poults may get lost when first introduced and never find feed and water. This, of course, can be overcome by good management, but will increase labour input. If a very big pen is built, it will almost certainly be of a semi-permanent nature. Sooner or later, disease problems will arise. One must then consider resiting the pen or relying on annual drug prevention or treatment programmes. It is clearly easier to re-erect smaller than larger pens.

In the author's view, a stocking rate of ten square yards (nine square metres) per poult is reasonably generous and will produce good quality pheasants provided that the pen contains plenty of cover and that the management is good.

B. STOCKING THE RELEASE PEN.

Any particular release pen should be stocked once only per season. Ideally, the stocking should be done at one time. If it is necessary to stock the pen in stages, ensure that as short a time as possible elapses between the initial introduction of poults and completion of stocking. Never attempt to mix poults from different sources in the same pen. The worst possible thing that one can do is to stock a pen with, say, 500 poults from one source and to stock it with a further 500 from a different source a week to ten days later. This is an ideal way to create an outbreak of disease. Suppose the first 500 poults were bought from game farm A and had been grass reared. Suppose that they were carrying Hexamita infection which was doing them no harm because of a combination of preventive drugs and immunity. These birds remain healthy but, nevertheless, seed the ground heavily with parasites. Game farm B then adds 500 poults which have not previously encountered this infection. It is wet weather and feed is scattered on the ground. In the first day or two, suboptimal amounts of feed and hence preventive drug are consumed, but large numbers of Hexamita are ingested. Within a week, over half the Game farm B poults may be dead. The keeper claims that the poults were not properly hardened off and demands compensation. In fact, the problem has entirely arisen because of the infection introduced from Game farm A combined with the keeper's own stocking policy.

C. AGE AT RELEASE.

It is unwise to release birds when they are less than six weeks of age since they are unlikely to be sufficiently hardy. If birds are released at much older than seven weeks, there may also be disadvantages unless one is releasing in small batches in roofed pens. The subjects of wing clipping and wing pulling were discussed in the previous chapter. Suffice it to state here that wing clipped poults will be capable of flying from the pen at approximately eight to nine weeks of age regardless of their age at release and regardless of when the wings were clipped. It follows that, when poults are released at between seven and eight weeks of age in open topped pens, they will generally spend one week less time in the pens than when released one week earlier. The only way to overcome this is to pull the wings. This technique has its own potential disadvantages. In general, the longer a poult remains in the release pen, the greater will be the likely return rate. It follows that, in the majority of cases, the best age to release is somewhere between six and seven weeks.

D. RELEASING COCKS ONLY.

It is sometimes recommended that estates which buy in poults should only purchase cock poults. This is particularly recommended for those who are trying to build up their stocks of wild birds and wish to shoot cocks only. On small shoots, the return rates for cocks are often up to 10 to 15 per cent better than for as hatched poults. This is because hens have a greater tendency to stray and to stray in larger groups. In many cases, therefore, it makes sense to release cocks only. However, if too many people adopt the practice, there is clearly going to be a problem with surplus hens. Those who want cocks only must be prepared to pay more for them than they would for as hatched poults so that those who take residual hens can pay less.

E. PROS AND CONS OF BEAK TRIMMING.

It is very common for birds in release pens to lose their tails through feather pecking. In an effort to avoid this, many people like to beak trim poults at release. Under no circumstances should beak trimming be severe. Even a minor beak trimming can reduce feed intake if feed pellets are scattered on the ground as is often the case in release pens. Failure of adequate feed intake at time of release can have a variety of very damaging consequences which will be discussed later. Minor beak trimming is unlikely to inhibit tail pulling for much longer than a week or ten days. Furthermore, tails are almost always properly grown out by the start of shooting whether they have been pecked in the pen or not. On balance, therefore, beak trimming should be avoided. If pecking problems other than tail pulling are so severe in the pen that beak trimming is considered essential, then there is almost certainly something badly wrong with the pen, its stocking rate or its management.

F. FEEDS AND FEEDING.
i) Quantity.

A typical pheasant at release will be consuming 45g/day. This will increase steadily to about 70g/day at 12 - 13 weeks and thereafter decline to 55-60g/day at 18 - 20 weeks of age. At this stage, the pheasant should have reached maximum body size and will subsequently consume approximately 400g per week. The quantities given relate to dry concentrate feeds such as pellets and wheat. During the period from release at six and a half weeks to 20 weeks of age, total consumption per bird will be in the region of 5.9kg. The proportion of this total that should be composed of pelleted feed will depend upon the protein level of the pellets. This subject will be covered in the subsequent section.

The above calculations are based on the assumption that the poults will be consuming no natural feed. For budget purposes, however, one can usually ignore the contribution from natural feed as this will tend to be counter balanced by the consumption by wild birds of feed offered to the reared pheasants. When special crops such as maize or quinoa are grown, they may make a much greater contribution to total feed intake from, say, fifteen weeks onwards.

Clearly, if one releases 1000 poults at six and a half weeks, one may be feeding considerably less of them by 20 weeks due to straying, predation and disease. The better one's keepering, the greater will be the numbers held and the higher the feed bill. It is extremely unwise of shoot owners to attempt to skimp on feed bills as an inadequate supply of feed or the provision of feed of poor quality will greatly increase straying. Keepers should be encouraged to keep careful logs of the quantities of feed put out each week. This will often give the most accurate available assessment of the actual numbers of birds on an estate at any stage. A reasonable budget figure for feed costs per bird released is £1.20, taking wheat at £120/tonne and pellets at £180 - £240/tonne, depending upon protein level.

The subject of flying ability in relation to feeding is a vexed one. Heavy pheasants which fly poorly are not necessarily or even usually the consequence of a faulty feeding regime. Bird weight is principally determined genetically and some game farms now produce large strains of pheasants which are unlikely to fly well from flat terrain. Equally, some game farm strains of pheasant are very docile and may choose not to lift on seeing a line of guns. This can be so even if they are not heavy. However, there is no doubt that, in theory at any rate, it is possible to influence weight and hence flying ability by feeding. Very poor feeding with inadequate protein intake can produce weak pheasants with poorly developed musculature. More likely is the development of excessive fat. Fat is deposited in the later stages of growth and during maturity when energy intake exceeds energy expenditure. Energy intake is most likely to be excessive when the feed offered is of high nutrient density and requires little effort to obtain. An experienced keeper who is hand feeding can effectively use a rationing programme that limits pellet or wheat intake to somewhat below the total requirement and forces birds to work for natural feed, which is normally of lower nutrient density, to make up the deficit. Where hoppers are used, either alone or in combination with hand feeding, no rationing is possible since hoppers should never be left empty. It should be noted that, of the concentrate feeds commonly offered to pheasants, maize provides approximately 8 per cent more energy per unit weight than wheat and wheat 8 per cent more than typical pellets. It follows that pellets are less likely to produce fat birds than is wheat, and maize more likely. However, it is generally uneconomic to feed pellets exclusively to older pheasants. Although wheat and maize are liked by pheasants, oats and barley, having lower energy levels, would often be more appropriate if one could be confident that the latter would not tempt one's birds to stray excessively.

ii) Quality

The usual feed for post release pheasants is a combination of some sort of compound pellet and wheat. Wheat has a protein level of 10 per cent, typical Grower 1 pellets of 24 per cent, Grower 2 pellets of 20 per cent and Covert pellets of 16 per cent. These levels and even the names ascribed to the particular rations vary from compounder to compounder. The protein level required to give an optimum growth rate falls with increasing age. At release, this should be slightly over 20 per cent and should decline steadily towards 10 per cent at maturity at 18 weeks. An excess of protein will do the bird no harm, but will be costly. A deficit will lead to reduced growth rates and inferior feathering. Eventually, the birds should catch up, but only early poults will have time to do this before the shooting season. By combining pellets and wheat in various ratios, a ration of appropriate protein level can be obtained. Table 1 demonstrates how this can be achieved with a variety of compound feeds and wheat. As can be seen from the Table, if one uses a higher protein pellet, wheat can be introduced earlier and form a higher proportion of the total ration. Although the cost of compound feed is greater the higher the protein level, this option is usually the cheapest and is always so if the wheat is home produced. The Table shows that, if one feeds a combination of Grower 1 pellets at 24 per cent protein and wheat, each bird taken from six and a half to twenty weeks will consume approximately 2.2kg pellets and 3.7kg wheat. If using a Grower 2 ration at 20 per cent protein with wheat, about 3kg pellets and 2.9kg wheat would be required. A combination of Grower 2 pellets, Covert pellets and wheat would use 1.1kg of the first feed, 3.2kg of the second and 1.5kg of wheat. This information has been used to construct Table 2 which can serve as a ready reckoner of compound feed requirements on the three feeding regimes for releases of various sizes. It should be re-emphasised that it has been assumed that poult losses post release will not be great and that the resultant expected decreased food consumption will be counterbalanced by consumption by wild birds and by

TABLE 1

Quantities (g) and proportions of wheat and a variety of compound feeds of differing protein levels to give optimum growth of pheasants from release to twenty weeks of age.

Age (weeks)	release - 8	9 - 10	11 - 12	13 - 14	15 - 16	17 - 18	19 - 20	Total
Feed Intake (g) per poult during each age period	500	825	975	975	910	875	830	5890
% Protein required in diet	20.5	18.5	17	15.5	14	13	10	
Proportion of 24% protein pellet and wheat to give correct protein %	75/25	60/40	50/50	40/60	30/70	20/80	0/100	
Quantity of 24% protein pellet and wheat needed in each period	375/125	495/330	487/488	390/585	273/637	175/700	0/830	2195/3695
Proportion of 20% protein pellet and wheat to give correct protein %	100/0	80/20	70/30	50/50	40/60	30/70	0/100	
Quantity of 20% protein pellet and wheat needed in each period	500/0	660/165	683/292	488/487	364/546	263/612	0/830	2958/2932
Proportion of 20% pellet, 16% pellet and wheat to give correct protein %	100/0/0	50/50/0	20/80/0	0/100/0	0/70/30	0/50/50	0/0/100	
Quantity of 20% pellet, 16% pellet and wheat needed in each period	500/0/0	413/412/0	195/780/0	0/975/0	0/637/273	0/438/437	0/0/830	1108/3242/1540

TABLE 2

Quantities of compound feeds of varying protein levels required to supplement wheat for pheasants from release to 20 weeks of age.

Numbers of Birds released	Quantity (Kg) and, in brackets, numbers of 25kg feed bags of 24% protein compound feed required as supplement to wheat	Quantity (Kg) and, in brackets, numbers of 25kg feed bags of 20% protein compound feed required as supplement to wheat	Quantities (Kg) and, in brackets, numbers of 25kg feed bags of 20% protein and 16% protein compound feeds as supplements to wheat
50	110 (4.4)	148 (5.9)	55/162 (2.2/6.5)
100	220 (8.8)	296 (11.8)	111/324 (4.4/13.0)
200	439 (17.6)	592 (23.7)	221/648 (8.8/25.9)
300	658 (26.3)	887 (35.5)	332/973 (13.3/38.9)
400	878 (35.1)	1183 (47.3)	443/1297 (17.7/51.9)
500	1097 (43.9)	1479 (59.2)	554/1621 (22.2/64.8)
600	1317 (52.7)	1775 (71.0)	665/1945 (26.6/77.8)
700	1536 (61.4)	2071 (82.8)	776/2269 (31.0/90.8)
800	1756 (70.2)	2366 (94.6)	886/2594 (35.4/103.8)
900	1975 (79.0)	2662 (106.5)	997/2918 (39.9/116.7)
1000	2195 (87.8)	2958 (118.3)	1108/3242 (44.3/129.7)

waste. Contribution from natural feed should be assumed to be replacing wheat and not compound feed.

Any one of the three recommended pellet/wheat combinations demonstrated in the Tables will provide adequate protein and energy to allow near maximum growth so that no compound feed is required from the nineteenth week. The Grower 1/wheat option will be the cheapest and the Grower 2/Covert pellet/wheat combination the most expensive. However, the difference should be no greater than 7.5% unless one has access to a cheap source of wheat. Ration cost is not the sole criterion when selecting an appropriate feeding regime. Most compound feeds fed post release contain preventive drugs, namely anti coccidials and anti blackhead medication. There are defined levels which feed compounders are allowed to include without recourse to a veterinary prescription. Thus, if one feeds a 24% protein feed pellet in combination with wheat in the release pen, each poult will be receiving less drug per day than would have been the case had no wheat been fed. Under certain circumstances, this could lead to an increased risk of disease. This subject will be discussed more fully in the subsequent section of this Chapter.

If possible, avoid feeding newly harvested wheat. Until it has been stored for a few weeks, it may sometimes precipitate digestive upsets and scouring.

iii) Drug inclusions.

As mentioned above, post release compound feeds often contain drugs which are included to prevent disease.

There are currently two anti coccidials that are licensed for use without veterinary prescription - lasalocid (Avatec) and clopidol (Coyden). The latter has been licensed for longer than the former and resistance to it has developed on some game farms and estates. Under ideal circumstances, one should be releasing into a relatively uncontaminated release pen and certainly not one that has had a previous release of pheasants through it earlier in the season. The birds released should have experienced a light coccidial challenge and the anti coccidials received should have prevented disease, but not been so effective as to have prevented the acquisition of active immunity. Under such ideal conditions, an anti-coccidial in the post release diet will not be necessary. However, on very rare occasions, released poults have had inadequate exposure to infection to allow immunity to develop and, if these are placed in a relatively contaminated pen, an outbreak of disease might develop in the absence of prophylactic (preventive) medication in the post release diet. In this situation, even a reduced daily drug intake, as would occur when diluting pellets with wheat, would suffice to prevent disease. More commonly, pheasants are released in which there is subclinical or even a low level of clinical disease. If their intestines contain relatively high levels of coccidia and drugs are suddenly withdrawn, disease is a likely consequence. As a general rule, therefore, post release diets should contain anti coccidial agents.

The anti blackhead drug, dimetridazole (Emtryl/Dazole), is also to be recommended in post release diets. While blackhead is relatively uncommon in pheasants, two other protozoal diseases, caused by *Trichomonas* and *Hexamita*, are much more frequent and can result in devastating losses. Dimetridazole has some effect in controlling these agents as well as *Histomonas meleagridis*, the protozoon causing blackhead. However, the level of active ingredient added by feed compounders is usually 125 parts per million (ppm). This level does not always appear to be adequate to prevent disease, especially when wheat is a supplementary feed. Compounders are allowed to add up to 200 ppm of dimetridazole (as Emtryl) to pheasant rations. If one anticipates a problem or proposes to dilute the pellets with wheat, one should request a higher drug inclusion rate. This will probably only be possible if one orders at least 2 tonnes of feed.

There are only two other drug which can be incorporated into compound feeds without veterinary prescription. These are mebendazole (Mebenvet) and flubenol (Flubenvet). The latter has largely replaced the former as it is cheaper and has a similar range of efficacy. Both drugs are wormers. The course of treatment with Mebenvet is designed to last for two weeks and with Flubenvet for one week. Clearly, if the compound feed is supplemented with grain, an inadequate quantity of drug will be consumed within the stipulated length of treatment. However, in the case of these drugs, efficacy can still be obtained if the treatment periods are extended by an amount which compensates for the degree of grain dilution. There may be complications when pheasants begin leaving the area of the pen and not returning to feed.

While it is recommended that post release diets fed up to 12 weeks of age should routinely contain an anti coccidial and an anti blackhead drug, there is no need to use a drug for gapeworm treatment as a routine in the feed although it may sometimes be convenient, albeit expensive, to do so. Gapeworm challenge may be so low as not to justify treatment. Even when treatment does appear necessary, alternative drugs exist which can either be added on to feed or given in water over a 24-hour period (Wormex and Gapex respectively).

iv) Feeding.

When pheasant poults are first put into release pens, it is essential that they are able to find feed and water immediately. One often observes that birds only eat half as much feed during the 48 hours immediately following release as they did in the equivalent time just before release. Many keepers will claim that this demonstrates that their pens contain plenty of natural feed. While this may sometimes be partly true, a more likely explanation is that a proportion of poults have failed to find feed pellets or water. This period is one of great stress for the poults and it is, therefore, imperative to attempt to minimise these drops in nutrient and fluid intakes. If birds are carrying subclinical disease, it is likely that a poor introduction to the release pen will trigger a problem, particularly as intakes of preventive drugs will be reduced. If one searches a release pen thoroughly at about two weeks after the introduction of poults, it is not uncommon to find corpses in the undergrowth, often amounting to 2 per cent of the release. These are usually birds that failed to adapt to the new conditions and which died through dehydration or through blocking their intestines with fibrous vegetable material because they had not found feed and water. The bulk of such mortality is preventable by good management.

Feeders and waterers are often sited in pens in such a way as to be most convenient to the keepers. Once the birds are established in the pen, this makes a certain amount of sense. In the immediate post release period, it is merely bad management. In the first week, feed should be placed throughout the pen, in troughs or piles or in hoppers which can be recognised by poults as feeders. Most of the feed hopper designs used post release will be unfamiliar to poults and they will require time in the pen before they learn to use them. If one has an exclusive hand feeding policy, under no circumstances introduce it straight away in the absence of alternative feeding points. Be prepared, if necessary, to waste a certain amount of feed at this stage. Never make the poults finish up the available feed before offering more.

Once the birds have adapted to the pen, one can settle down to a more routine feeding strategy and can opt either for a policy of hand feeding, hopper feeding or a combination of both. If hopper feeding, the birds must learn over a period of time that the hoppers are a source of feed (*fig 18*). Feed may be obtained by pecking at wire mesh, slits, pendulums, or through 'letter boxes' depending upon design. All can be effective, but some deliver feed more quickly than others. There should be a concentration of hoppers in the pen initially and a few immediately outside for the benefit of premature escapees which may take time to learn to use the popholes efficiently. As birds start moving from the pen, some of the hoppers should move with them. Eventually, there may be groups of hoppers in each game cover or drive and very few in the region of the pen. In the intervening period, it is often worth while to space hoppers along hedges that lead to game covers, thus establishing in the minds of the pheasants which are desired routes of travel between feeding and roosting areas. For this reason, a proportion of hoppers should be readily mobile and one should not rely exclusively on those that are anchored to the ground by stakes.

It must be appreciated that one large hopper is very much less effective at holding birds than several small hoppers. At older ages, hens form quite large feeding groups which are well adapted to feeding from either strawed feed rides if hand feeding is practised or from reasonably closely sited groups of hoppers. Cocks feed in very small groups and are aggressive. Large numbers of spaced hoppers are thus best suited to their behaviour.

Fig. 18
Releasing poults. When feeding poults in the release pen they should initially receive feed from feeders to which they have previously been used to, but should also be introduced to the hopper types that they are subsequently going to encounter on the estate.

A reasonable compromise is to aim to provide one hopper per 15 to 20 birds released. As these hoppers will eventually be scattered throughout the estate, it is sensible to establish strategically sited depots of feed at a time when vehicular access is easy - after harvest and before the ground is cultivated.

While hopper feeding is less labour intensive than hand feeding and tends to prevent pheasants from becoming too tame, it does have disadvantages. The birds are less manageable. They can feed quite quickly and then either return to their roosting areas or wander excessively and increase labour requirements for dogging in. One can rarely be sure to find them in a particular game cover at a particular time. This is the great advantage of hand feeding provided that it is assiduously carried out every day, several times a day and at regular times. Another advantage is that a good keeper can ration his birds and prevent excessive fatness. To be really effective, however, a great degree of commitment, hard work and skill are required. When hand feeding, feed losses to song birds will tend to be higher than with hopper feeding.

Scattering grain on to a clean, mould free strawed ride is how one normally envisages hand feeding. This poses few problems. However, when first introducing hand feeding in the release pen, the scattering of pellets on to bare ground is more usual. Under wet conditions, particularly, this can cause problems as the ground will become quickly fouled and the pellets will disintegrate. Poults will inevitably peck up a lot of soil and faecal contaminants while feeding and will be liable to protozoal and worm infections. One should attempt to feed in drier areas of the pen and to move on to new ground every five or six days. Even when one relies very heavily on hand feeding, there is a case to be made for using hoppers in outlying areas of an estate which cannot be regularly visited.

Several types of battery operated automatic feeders are available which scatter feed over a limited area. These can be adjusted to provide several feeds per day of varying magnitude. They emit a sound prior to feeding to attract the pheasants. In theory, these may seem to be very useful devices. However, they have several disadvantages. They are expensive and one feeder cannot be expected to feed more than a maximum of one hundred poults, given the diameter of the feeding area in relation to feed competition and aggression on the part of the birds. The sound they emit is not audible over long distances, despite the claims of some manufacturers. There is no scientific evidence that pheasants can hear better than humans at any particular sound frequency. The battery and machine itself often require attention or adjustment. Most designs have photosensitive cells which switch them off at night. The first feed is usually provided at dawn. Subsequent feeds can be provided at desired, but regular, intervals thereafter until dusk. This is not what is normally required. As pheasants leave the pen, one usually wants them to go and feed in areas from which they will eventually be driven. One may then wish to feed them in their release area to which one hopes they will return to roost in the evening or late afternoon. An automatic feeder, with a photoreceptor, at this stage, can therefore only be used for the provision of one feed per day.

Maize is a commonly grown cover crop for pheasants and the grain is a very useful source of feed which can potentially save the purchase of a proportion of the wheat that would otherwise have to be fed. However, if one examines an average crop of maize towards the end of the season, it is surprising how many of the cobs remain intact. It is worth taking cobs to the release pen area and peeling back the leaves in the early part of the season. The released birds will learn more quickly to make use of the maize in this way. Once they are using the crop, one should ensure that cobs are regularly knocked down within reach throughout the season. However, it should be appreciated that maize grain has a very high energy content and can predispose to fatness.

Some compounders manufacture feed blocks for pheasants. One hears claims that these are attractive and prevent straying. Pheasants have very limited senses of taste and smell. They normally prefer pelleted feed to whole grain and, as the feed block is more akin in formulation to pellets than to grain, it is probably fair to claim that it will be preferred to wheat provided that it does not immediately disintegrate if not sufficiently protected from the weather and also provided that it has not been made so dense that birds cannot peck from it. However, since a given amount of nutrition from a block will usually cost at least double that from a compound pellet, there is little or nothing to commend the block. If one hopes to prevent excessive straying by providing feed that is more attractive than wheat, use pellets in a hopper.

G. WATER.

When discussing feeding, the importance of ensuring that newly stocked poults could immediately find feed was stressed. This is even more important with respect to water. While starvation takes a long time to kill birds, death through dehydration is rapid. The provision of half a dozen drinkers in the centre of a pen of an acre or so is simply not adequate. Such a line of automatically supplied drinkers may suffice after a week or so, but, initially, there should be large numbers of open water dishes or fonts throughout the pen with emphasis placed on the periphery since newly released poults have a tendency to walk the perimeter wire. There should also be water immediately outside the pen for premature escapees. Long term, open water pans with no reservoirs are unsuitable. Not only are they quickly emptied, but they are rapidly fouled as poults climb into them and deposit mud and faeces. The contaminated water is an ideal breeding ground for bacteria and other infective agents. Poults may thus become diseased or, alternatively, reject the water and grow poorly as a result. Although it has previously been stressed that pheasants have little sense of taste, they actively dislike rusty water.

If there are several drinkers in a pen and some are empty, it should not be assumed that all poults will wander off to find one that still contains water. Some, having learnt to use one drinker only, will wait in its vicinity until it is refilled. While waiting, they will not feed adequately. When the water is replenished, they may over consume it. This may not always matter, but, if one happens to be medicating through the water, excessive drug intake can occur with adverse consequences. It is thus better to remove drinkers altogether than to leave them empty.

With large releases, it is almost essential to arrange for some degree of automation of drinking supplies - at least in the release areas. This would normally be provided by gravity fed drinkers supplied through raised water reservoirs. The latter can be filled from a water bowser. It is surprising how often a keeper's time is wasted by the constant and heavy chore of hand carting water. Time is thus lost from other duties such as dogging in. More important, the pheasants often go short of water, particularly in dry seasons. Although this may very occasionally be so severe as to cause death due to dehydration, the usual manifestations of water deprivation are more subtle. These include poor growth, increased tail pulling and straying.

If pheasants drink from stagnant water bodies in which the vegetation suddenly dies off or is cut and left to rot, there is a risk of botulism which can kill very large numbers of birds. For obvious reasons, however, the problem is much commoner in ducks than in other gamebirds.

The approximate daily water consumption of pheasants from 3 to 12 weeks is summarised in a graph (*fig 19*). The solid line relates to quantity drunk by each bird each day. The dotted line relates to volume drunk per kg body weight of pheasant per day. The latter can be used to calculate the quantity of medication to add to water if drug treatment through water becomes necessary. As an example, if one wishes to treat 6 week old poults and the instructions with the drug state that birds should be treated at the rate of 10mg/kg/day, one can see from the graph that 175ml of water/kg pheasant/day will be drunk. This means that one should dose the water at the rate of 10mg per 175ml. Therefore, each litre (1000ml) of water should contain (1000 divided by 175) x 10mg = 57mg. If the drug is in a solution containing 20mg of active ingredient per ml, one would achieve one's daily desired dosage by giving 57/20 = 2.9ml of drug per litre of water. The graph cannot be regarded as being other than an indication of water consumption. If birds are eating a lot of green stuff, they will drink less. Dietary protein level affects water intake markedly. The higher the level at any age, the higher will be the water consumption at that age. Hot weather increases water intake, but not enormously unless it is so hot that birds are panting a lot. Scouring, as may be caused by protozoal infections, can lead to very large increases in water consumption.

H. PROBLEMS AND THEIR PREVENTION.

i) Failure to find feed and water

As was stated in a previous section, failure to find water quickly after release can lead to rapid mortality. In dry seasons in particular, losses can be quite significant. This is a management problem that can be overcome by the provision of adequate numbers of permanently filled drinkers within the pen and the placing of some outside it to cater for premature escapees.

Failure to find feed is less likely to kill poults directly, but can precipitate a variety of disease problems which themselves can be fatal.

ii) Exposure and heaping.

Cold, wet and windy weather coinciding with release can cause major losses. If poults are well feathered, healthy and have been well hardened off, problems are less likely. A properly conditioned poult will have received no artificial heat within the ten days prior to release and should have been well wetted two or three times either by rain or by spraying with a device such as a knapsack sprayer.

The pen itself should contain plenty of cover to give protection against draughts and partial protection against rain, but there should also be open areas to allow birds to dry out after rain. Small roofed areas within the pen are

Daily water consumption of pheasants

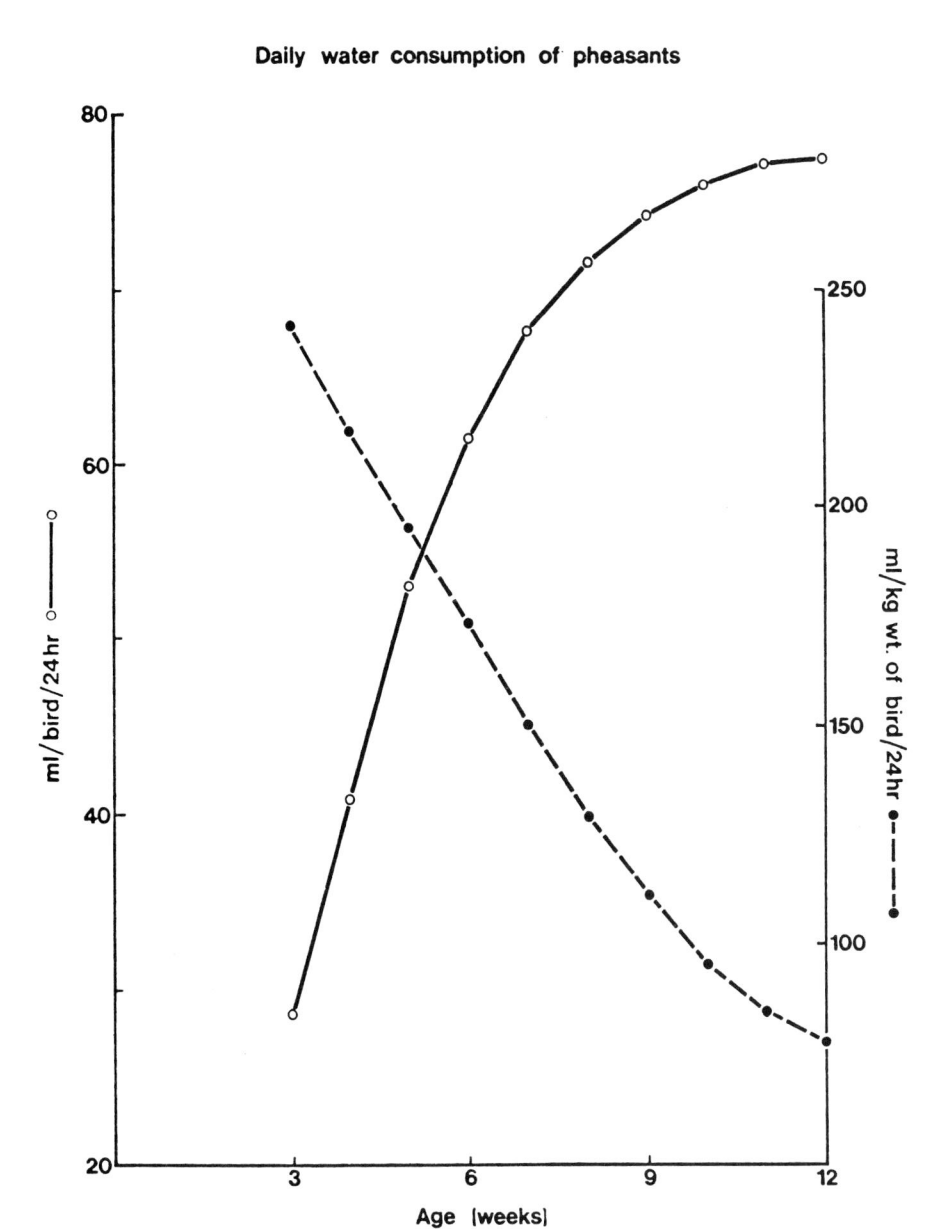

Fig. 19
Average pheasant water consumption expressed in volumes per bird per day and volumes per weight of bird (Kg) per day.

often useful in providing extra protection both for birds and for feed hoppers or troughs. While these should not have sides, they should nevertheless be free from draughts or they will not be well used.

It is best to release pheasants earlier rather than later in the day so that they have several daylight hours to explore their new environment. Obviously, one should listen to weather forecasts and not release birds when continuous rain is anticipated. Short sharp showers are not particularly damaging while prolonged cold rain can be. Once poults have been released for 72 hours, bad weather will not directly harm them although very wet conditions can precipitate disease problems and feather pecking.

iii) Gut impactions

As soon as poults are released, they will start to peck at any objects that attract them. If a bird pecks at a large piece of grass which has been cut and is, therefore, not anchored to anything, it will start to swallow it. It cannot regurgitate and is thus committed to swallowing the entire piece. If it does this several times, its gizzard will become impacted with a ball of fibrous material which will prevent the ingestion of proper feed. The bird may slowly die, or, if the impaction is not complete or gradually unblocks, will merely become weak and emaciated. If the grass is growing when the poult pecks it, the problem does not arise as each peck will break off a bite sized chunk from the blade of grass. One should, therefore, never cut long vegetation in the pen and leave it lying around. It should be raked up and removed. This will also reduce the risk of aspergillosis due to moulding of the cut material.

Lengths of string and binder twine can also cause impactions as well as getting wrapped round legs. Poults will peck at shiny objects and dropped nails and staples are often swallowed. These often penetrate the gizzards and cause deaths from peritonitis.

If there is little or no natural insoluble grit in the release pen, it is sensible to scatter some on the feed once every week or ten days. This will enable the gizzard to grind up fibrous material.

iv) Tail pulling and vent pecking

The temporary loss of tails through pecking is an extremely common problem post release. The more birds there are in a pen, the worse the tail pulling tends to be. This is because, even in large pens, poults tend to be concentrated at high densities in certain areas. In very wet conditions, in which poults tend to group together in miserable heaps, the vice can become severe and can progress to vent pecking. Once vents are damaged, the affected poults will tend to lose condition and die. Under most circumstances, the tails regrow satisfactorily before the shooting season and one thus sees what is, effectively, a delayed moult. In dry seasons, when free water and good quality food are limited in areas away from the release pen, the regrowth may be very slow indeed. This is probably because the pheasants, having left the pen, have to remain within its vicinity at high density in order to find food and water. Having acquired the pecking habit, they continue to indulge in it. The practice of hand feeding, which tends to encourage large numbers of pheasants to congregate in small areas either waiting to be fed or actually feeding, exacerbates the tail pulling problem. Hopper fed birds can feed quickly and thereafter keep out of each other's way.

The subject of beak tipping has been previously discussed. It is more likely to delay than to prevent tail pulling, but, worse, can inhibit the ability of poults to pick up pellets from the ground when they are first released.

The best way to minimise the problem is to release pheasants into pens with plenty of ground cover and ensure also that there is a lot of low roosting cover accessible to wing clipped poults. Thereafter, be generous with drinkers and feed in such a way as to ensure that poults are well dispersed throughout the pen and do not remain concentrated in small parts of it. Once birds start leaving the pen, make sure that feed and water are reasonably dispersed around the estate.

Ensure that the diet offered is not deficient in protein. Tail pulling is usually worse on estates which feed too little in the way of pellets and too much wheat. A lack of salt can increase aggression and feather pecking. Wheat is salt deficient and some keepers claim to prevent problems by putting lumps of rock salt into pens. These should obviously be protected from rain. Apparently, poults readily peck at them and are less likely in consequence to pull tails.

v) Worms

By far the most important helminth (worm) parasite of pheasants in the release pen is *Syngamus trachea* - the gape worm (fig 20). The adult parasites live in the trachea (wind pipe) of infected birds. The larger females remain attached to the smaller male worms in permanent copulation while the mouthparts of each worm are buried in the tracheal wall. The worms are red in colour due to pigmentation and not to blood sucking as is sometimes believed. They are easily visible to the naked eye if the trachea is split open at post mortem examination. The effect of the worms is to cause respiratory distress with snicking, coughing and gaping. The severity will depend on the diameter of the trachea in relation to the size, numbers and distribution of worms since it is physical obstruction of the airway that creates the problem. If very severe, suffocation and death will ensue. If less severe, growth will be checked and birds will be more vulnerable to other diseases and to predation. If there are only a few, well spaced out pairs of worms in the trachea of well grown poults, there may only be snicking followed, after approximately two weeks, by self cure due to the establishment of immunity and the expulsion of the worms. The resulting birds will be resistant to re-infection until the following breeding season at which time the immunity will tend to break down.

PHEASANT HEALTH AND WELFARE

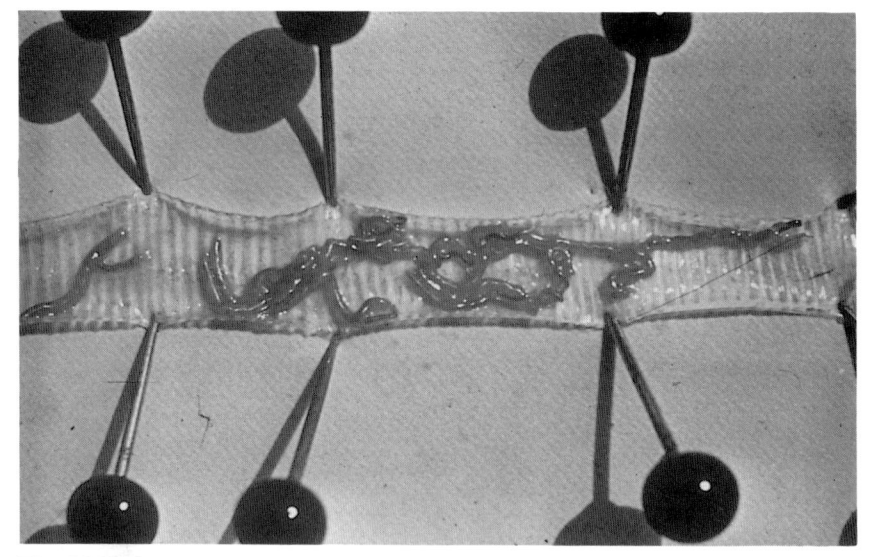

Fig. 20
Syngamus trachea *(gapeworms) in the trachea (windpipe). The small male worms are permanently attached in copulation to the larger females. Death can be caused by asphyxiation.*

Adult female gapeworms lay eggs that are coughed up and then swallowed by the birds. After they have passed out in the faeces, the eggs are not immediately infective. First, they have to develop into larvae which may remain for some time within the egg shell. Under favourable conditions - warm and damp - this development can occur within five days. If larvae are swallowed by poults they penetrate the gut wall, migrate through liver and lungs and arrive in the trachea within four days. Within eight days, there may be signs of respiratory distress and, from the thirteenth day, the worms will be producing eggs and starting another cycle of infection. Many larvae will not be immediately eaten by birds and most will die over a prolonged period. A few, however, will be swallowed by earthworms or other invertebrates. The larvae will remain dormant in these. Should they be eaten in turn by pheasants, the larvae within them will then develop into adult worms. Earthworms can live for six or seven years. Once a pen is infected, therefore, it will remain so even if given a year's rest. It is not only totally impractical, but also environmentally damaging to attempt to eliminate earthworms from a release pen. In any event, earthworms and other invertebrates, are not major sources of infection. Many larvae survive in the environment for long periods. If adult pheasants have access to the pen a few months before poults are stocked into it, it will probably be contaminated. Equally, pigeons, rooks, blackbirds and several other species of wild bird carry gapeworm infections and can seed release pens with eggs.

POST RELEASE MANAGEMENT

Having discussed the life cycle of the gapeworm, one can proceed to consider its control. Clearly, there is unlikely to be a major problem in relatively new release pens. However, even in well established release pens, heavy infections requiring treatment are not inevitable. In very dry conditions, for example, the eggs do not develop well. Even when they do, it is not inevitable that large numbers of larvae will be eaten by the poults. If, for example, the feed is kept off the ground, the level of infection will be greatly reduced. The practice of hand feeding on bare, damp ground offers the best possible opportunity for the problem to develop. If the pen has been stocked in the same season with a previous release of birds, one should expect a much more severe infection. However, as has been stated previously, this is a totally unacceptable management practice which is likely to precipitate a whole range of problems some of which will have more serious consequences than gapes.

There are four non-prescription drugs which may legally be used to control Syngamus. Three are related benzimidazole drugs. Benzimidazoles have a wide spectrum of activity and will thus kill several worm species. Furthermore, they kill immature as well as mature worms. They are not soluble in water and have to be administered via the feed - either incorporated into it by the feed compounder or added on to it by the keeper. The three drugs are Mebenvet, Flubenvet and Wormex. Mebenvet is a powder which can be given in or on feed, Flubenvet has to be mixed into feed by a compounder and Wormex is an oily suspension which can only be given on feed. The recommended course of treatment with Mebenvet is two weeks, with Flubenvet one week while that for Wormex is one to three days, depending upon the keeper's wishes. The fourth drug, Gapex, is given via the drinking water over a twenty four hour period. Gapex, while highly effective against adult worms, has little effect against immatures. Unlike the other drugs, it is potentially toxic and must be used with care, precisely according to the manufacturer's instructions.

Timing of treatment is important. Some keepers, for example, buy post release feed pellets containing Flubenvet or Mebenvet as a routine in the knowledge that they always experience gapes. If they start to feed these as soon as the birds arrive, the poults will not become infected during the periods in which the drugs are being fed. However, as soon as the drugs are withdrawn, the poults will be totally vulnerable to infection. Admittedly, they will be one to two weeks older and thus somewhat better able to withstand its effects. If the medicated feed is not given until the birds have been in the pen for two weeks, the majority of the poults will, by then, have probably picked up infection. A course of treatment at this stage will kill the worms and leave the poults resistant to re-infection. However, in these circumstances, a proportion of the poults might have died before the drug has had a chance to kill the worms and others may have left the pen and not be reliably returning for medicated feed so that they never complete the course of treatment. Furthermore, more feed and hence more expense will be incurred by treating older rather than younger poults. A reasonable compromise, therefore, is to give medicated feed one week after release.

When using Wormex, keepers can use more discretion since the course of treatment is so much shorter. They will obtain the best results by waiting until significant numbers of birds - perhaps one quarter of the total - have started to show signs. If they treat at this stage, most of the poults will be infected and thus immunity will have been stimulated, making re-infection much less likely. When using Gapex, one should, if possible, delay for even longer since the drug is only effective against worms that have been in the birds for at least eight days. Its effect can then be highly dramatic in that the signs of disease cease abruptly and the birds are fairly solidly immune. Water treated with Gapex is not particularly palatable so that it is less likely to be effective in wet conditions or if there are natural sources of water within the pen.

It is well established in other species that worms can become resistant to benzimidazoles. There have been reports of gapeworm resistance to benzimidazoles in pheasants, but these have not been substantiated scientifically. However, one can reasonably anticipate that the problem will sooner or later occur, particularly if wormers are used routinely and indiscriminately. It should be appreciated that resistance to one benzimidazole usually results in resistance to all benzimidazoles and that, other than Gapex, there are no alternative candidate drugs which are likely to be effective against gapeworms. One way of delaying or preventing the onset of resistance on a particular estate, which is obliged to treat annually, is to use a benzimidazole drug one season and a non benzimidazole drug the next.

Various species of Capillaria worms and Heterakis can also affect pheasants in release pens. However, problems are very much less common than those associated with gapes. While it is only 13 days between the ingestion of gapeworm larvae to the time that gapeworm eggs are being produced, the interval with Heterakis and Capillaria species is much longer. It is thus unlikely that high levels of infection will be picked up in the release pen unless management has been very poor.

Heterakis worms are to be found in the caeca or blind guts of birds where they do relatively little harm unless present in large numbers. They can be seen by naked eye at post mortem examination. Different species of Capillaria occupy different positions throughout the length of the gut. Although quite long, they are very thin and tend to be difficult or impossible to see. Those infecting the crop tend to be most damaging. A veterinary diagnosis will almost certainly be needed in cases involving Capillaria infections.

Heterakis are very easy to destroy with benzimidazole drugs, but some species of Capillaria, particularly those in the crop, are not nearly so easily killed. Higher than normal doses may be necessary to obtain really useful effects. Flubenvet and Mebenvet are the only drugs licensed for treatment of worms other than gapeworms in pheasants.

vi) Protozoal Infections.

The diseases relevant to this section are coccidiosis, blackhead, hexamitiasis and trichomoniasis. All involve the intestinal tract, but blackhead can also damage the liver.

a. Coccidiosis.

This problem, or group of problems, is seen more commonly during rearing than in release pens and is discussed more fully in the previous chapter and also in Section F(iii) of this chapter. If one does experience coccidiosis in the release pen, it will indicate either the failure to use any anti coccidial, the use of an anti coccidial drug to which there is resistance, a highly contaminated release pen due to previous stocking the same season or to the release of birds already suffering from coccidiosis. However, if birds become sick with another disease or can't find feed, they may develop secondary coccidiosis, partly because what immunity they have tends to break down and partly because lack of feed intake results in lack of anti coccidial intake. If one is purchasing poults from a game farm, one should determine in advance what anti coccidial drug was used by the farm and, more important, whether it has a resistance problem with one of the two possible anti coccidials. If it has, there is no point in using that particular drug in one's post release diet.

b. Blackhead.

This disease is caused by *Histomonas meleagridis* and is uncommon in pheasants. If it does occur, one can experience sudden deaths of poults in good condition. The caeca (blind guts) are severely damaged, and, in birds taking longer to die, there are circular, yellow lesions on the liver surface.

Histomonas organisms passed in the faeces have very little ability to survive in the environment unless they are protected within the eggs of the caecal worm, Heterakis. Heterakis eggs are themselves quite tough and persist on the soil for long periods. Furthermore, they can, in turn, be ingested by earthworms and survive in them for even longer.

Pheasants are less susceptible to blackhead than turkeys and partridges and, in any event, an anti blackhead drug is normally incorporated into post release diets. Should blackhead occur on an estate, consideration should be given to the Heterakis status of the stock as it is likely to be unusually high. Blackhead can be treated with dimetridazole in the water. Although this is the same drug as used for prevention, it is used at a higher level and requires veterinary prescription.

c. Trichomoniasis and Hexamitiasis.

These diseases, caused by *Trichomonas phasioni* and *Hexamita meleagridis*, cause identical signs, although the latter tends to be the more serious. Both agents are small highly mobile flagellated protozoa which actively swim in the gut fluid of the small intestines. Diagnosis is greatly eased if live birds are submitted for post mortem examination as the swimming activity ceases very soon after death. There are no obvious gross lesions to be seen at post mortem examination except for a somewhat fluid-filled dilated small intestine containing very large numbers of flagellates. Affected birds lose weight extremely rapidly and walk about in a stilted and dazed manner. They continue to feed until shortly before death, but they scour badly, fail to digest the feed adequately and dehydrate. Mortality can be very high - up to 80 per cent in some cases of hexamitiasis. Outbreaks typically occur in poults of between seven and ten weeks although they can occur earlier or later.

Neither disease is well researched, but both are increasingly being diagnosed in pheasants, particularly when releasing in wet seasons. It is known that both agents can be carried in the intestines of healthy birds without causing disease. When trichomonads are shed in the faeces, they survive poorly in the environment, particularly in dry conditions. Hexamita is somewhat tougher. It is considered that Hexamita requires to be passed from bird to bird before it starts to become damaging. In other words, if it quickly goes through several generations, it hots up and causes disease. Problems are, therefore, more likely to be encountered if there have been multiple releases through one pen or if large numbers are released into a single pen. Grass reared poults are far more likely to be carriers of infection than those reared under cover although they, themselves, may remain healthy.

The chances of both infections are greatly increased if feeding and watering arrangements are unhygienic. Open water pans in which birds can tread, defaecate and drink are ideal for the survival and spread of these agents. Distribution of feed on to bare, damp ground which will also certainly be contaminated with faeces is the other prime way of creating these diseases. If one is hand feeding in a release pen, there is a lot to be said for distributing the feed in gutters or troughs in the early weeks so that feed and faeces are separated. If this is not practicable, thought should be given as to how feed is offered. It should not always be scattered over the same localised area which will rapidly become contaminated.

As an aid to the prevention of these two diseases, one should incorporate dimetridazole into post release diets. However, one should not rely upon the efficacy of this approach and neglect the need for good husbandry. As was stated previously, knowledge of these diseases is limited and this applies to drug therapy as much as to methods of spread. There have been many cases of hexamitiasis or trichomoniasis in birds which were receiving dimetridazole at a level of 125ppm in the diet. Some veterinarians recommend levels in the order of 200ppm.

Treatment must be undertaken through drinking water with drugs supplied on veterinary prescription. There is a limited choice of drugs of which one is dimetridazole administered at a level higher than is used in the feed for prevention.

vii) Marble Spleen Disease (MSD).

This disease is caused by an adenovirus which can also infect chickens and turkeys. It is probable that the vast majority of British pheasants have, at some time in their lives, been infected with this virus. Only in a small minority of cases does infection cause disease or death. The reasons for this are fairly well understood. To start with, there are several strains of the virus, some of which are very mild in their effects and a few of which are very pathogenic or damaging. Infection with a mild strain will induce immunity which will protect against a pathogenic strain. In the United States, mild strains are used as vaccines administered through drinking water. There is currently no licensed vaccine in the UK.

If a hen pheasant is immune, her progeny will receive passive immunity which will protect them for four to six weeks. If they are infected with either a mild or pathogenic strain during this period, they will develop active immunity and never subsequently develop MSD. It is only if they lose their passive immunity and are thereafter infected with a pathogenic strain that disease will occur. Even then, only a proportion of the birds will die. Stress, such as a severe fright, often appears to precipitate an outbreak.

The disease is characterised by sudden death of birds in good condition. There are no warning signs. At post mortem examination, the spleens can be seen to be somewhat enlarged, but the main feature -that which actually kills - is the congestion and oedema of the lungs. The poults drown in the fluid that is suddenly leaked into their lungs. It is more usual to encounter outbreaks after poults have left the release pens, possibly due to the fact that they are then more likely to mix with other stocks of birds with different histories of viral exposure.

MSD, being a viral disease, is not treatable. Although, in theory, if a totally non immune group meets a pathogenic strain of virus, mortality may reach 80 per cent, this very rarely occurs. Losses are much more frequently below 10 per cent. Due to the prevalence of infection and lack of a vaccine, there is little or nothing practical that can be done in management terms to minimise the risk of this disease.

viii) Mycoplasmosis.

This problem has been discussed in the chapters dealing with egg production and rearing. Suffice it to repeat here that it is a chronic, long lasting disease that will debilitate birds without killing many of them. It is characterised by coughing and by sinusitis below the eyes. The sinus swellings may cause the eyes to close, effectively blinding the birds.

Every effort should be made not to release poults suffering from this infection unless it is known that the infection is already endemic on the estate. If the problem is only noticed after release, it is best to shoot the stock out during the shooting season and replace it from a different source. Do not catch up hens from infected estates for breeding because the infection can be transmitted through eggs.

Prescription medicines are available which will control the clinical signs of the disease, but the infection will remain in the birds for as long as they live. Apart from spreading through the egg, mycoplasmas are passed from bird to bird by direct contact. However, contact has to be close as mycoplasmas do not survive well in the environment and are not spread through the air as are many respiratory disease agents. Therefore, if one does experience a problem in the release pen, it is unlikely to spread very much to other poults before they disperse from the pen.

ix) Undersized, poorly moulted birds at start of shooting season.

It is by no means unusual in mid October to encounter pheasants which look as though they will be too small and poorly feathered to shoot by mid November. Such birds may well have been released in July. Equally, it is possible, on other estates, to see mid August released pheasants which will obviously be ready to shoot at the beginning of November.

Chronic disease may check growth and feathering. More usually, the cause is either lack of dietary protein or lack of water with resulting lowered feed intake. Low dietary protein will almost always be due to the feeding of too much wheat at the expense of pellets.

There is no doubt that a pheasant of seventeen weeks of age that has been well managed and correctly fed should be in mature plumage with a tail of at least three quarters length. It should also have reached over 90 per cent of its fat free body weight. In other words, further weight gain will be mainly fat. However, this is not to imply that it will, by this age, be able to fly at peak performance, mainly because it is probable that this requires the practice gained by a few earlier expeditions over guns.

x) Summary of Problems

Obviously, not all possible problems have been covered. For example, no mention has been made of the viral disease, fowl pest, or bacterial septicaemia due to erysipelas infection. However, because common things commonly occur, the previous nine subsections will probably cover most of the problems that one is likely to encounter. As will be appreciated, the vast majority are avoidable provided that one starts with healthy, well hardened poults and thereafter manages them correctly. An exception is Marble Spleen Disease. Good management, of course, implies that a diet is fed which is correct not only in terms of protein, but also in terms of prophylactic medication.

POST RELEASE MANAGEMENT

An obvious omission from this chapter is the failure to discuss predators and predator control. However, this text has been written as a manual of pheasant production and, as such, concerns the 'poultryman' role of a keeper who is involved with raising pheasants. It makes no pretence of dealing with his several other equally important duties.

ID# APPENDIX 1
Anatomy

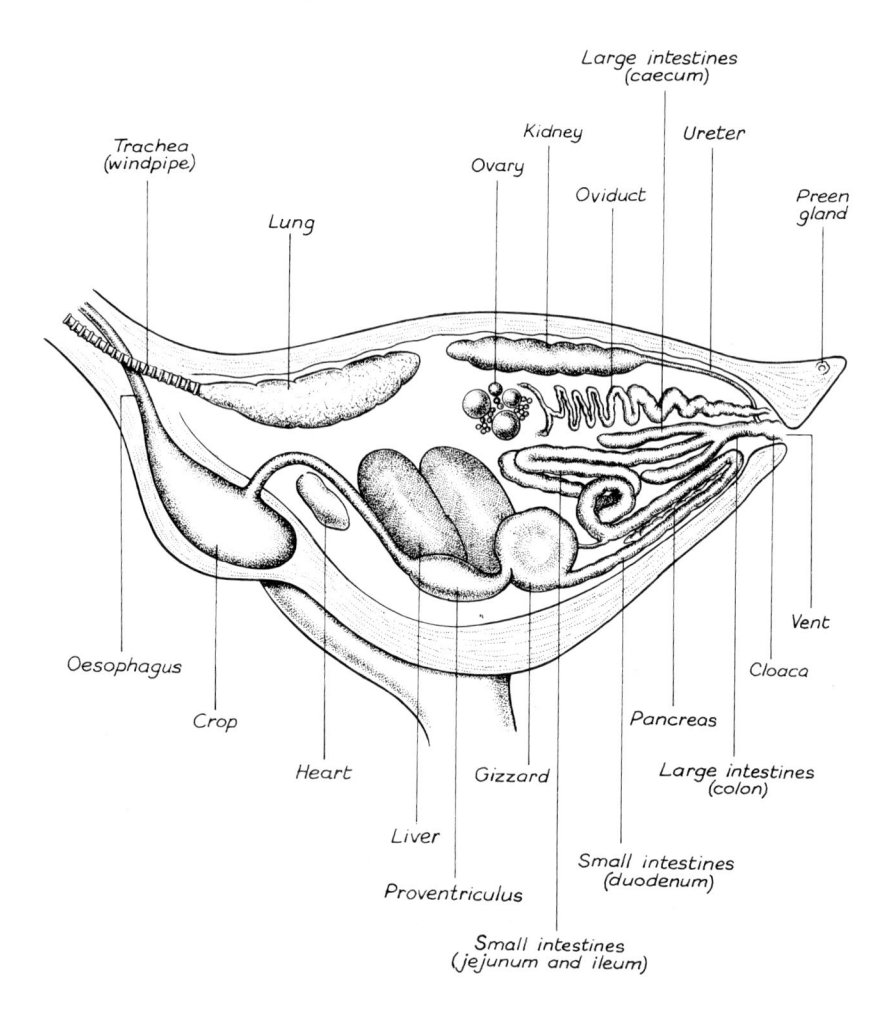

Fig. 21
Section through the body of a hen pheasant showing the layout of the digestive tract and viscera (air sacs omitted). *(Courtesy of J. Fuller)*

APPENDIX 1

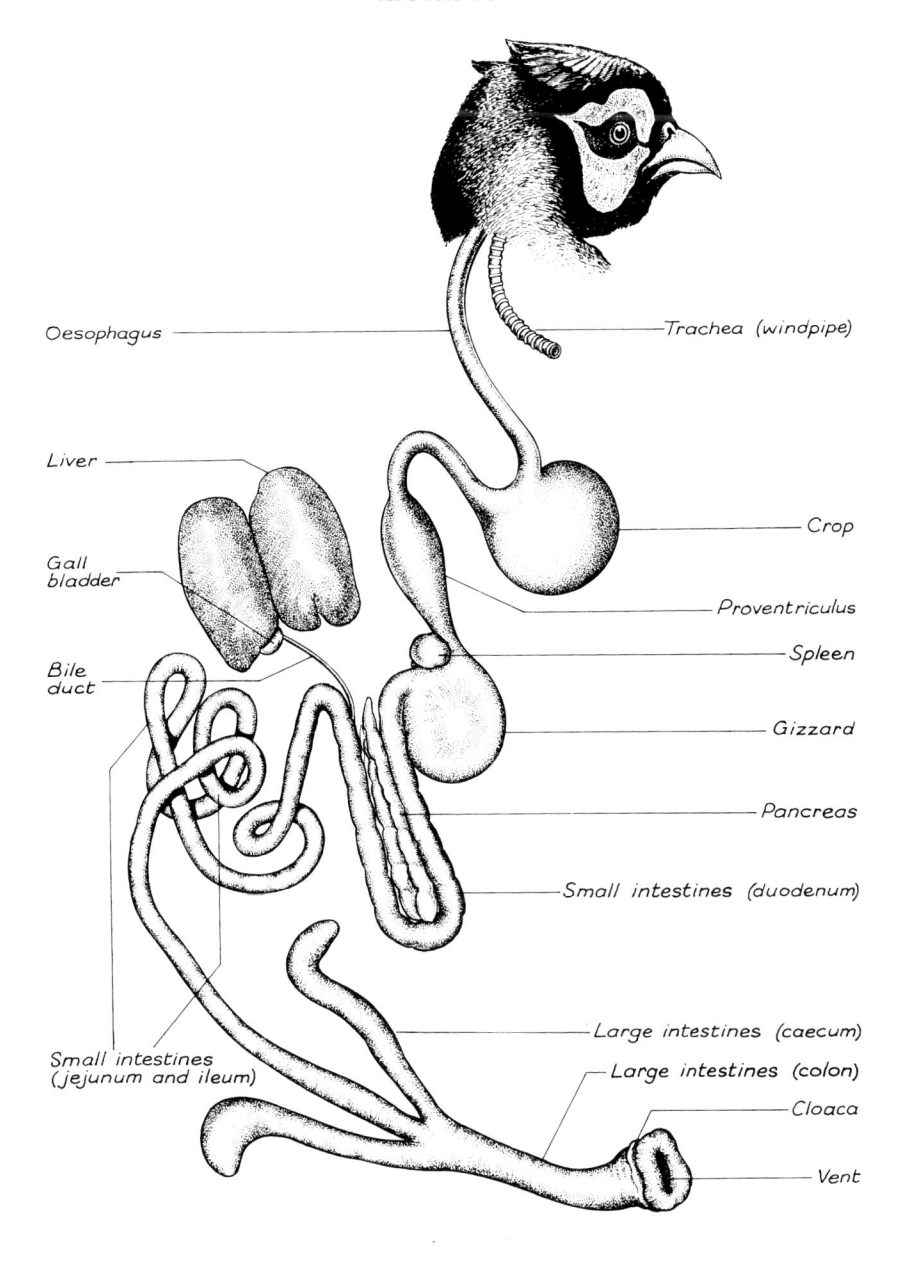

Fig. 21
The digestive tract of a pheasant.   (Courtesy of J. Fuller)

APPENDIX 2
Conversion Tables

Weights:	454 grams (g)	=	1 pound (lb)
	1000g	=	1 kilogram (Kg)
	1 Kg	=	2.2 lbs

Lengths:	2.5 centimetres (cm)	=	1 inch (in)
	100 cm	=	1 metre (m)
	12 in	=	1 foot (ft)
	3 ft	=	1 yard (yd)
	1760 yd	=	1 mile
	1000m	=	1 kilometre (Km)
	1600m	=	1 mile

Areas:	9 square ft	=	1 square yd
	10 square ft	=	1 square m
	1 acre	=	4840 square yd
	1 hectare	=	10000 square m
	1 hectare	=	2.5 acres
	1 square mile	=	640 acres
	1 square Km	=	100 hectares
	1 square Km	=	250 acres

Volumes:	1 millilitre (ml)	=	1 cubic centimetre (cc)
	1000 ml	=	1 litre (l)
	4.55 l	=	1 gallon
	8 pints	=	1 gallon
	1 pint	=	568 ml
	1 fluid ounce	=	28 ml